OTHER COUNCIL FIRES
WERE HERE BEFORE OURS

OTHER COUNCIL FIRES WERE HERE BEFORE OURS

A Classic Native American Creation
Story as Retold by a Seneca Elder, Twylah Nitsch,
and Her Granddaughter, Jamie Sams

*The Medicine Stone Speaks from
the Past to Our Future*

Jamie Sams and Twylah Nitsch

HarperSanFrancisco
A Division of HarperCollins*Publishers*

Illustrations by Linda Childers

FIRST EDITION

Library of Congress Cataloging-in-Publication Data

Sams, Jamie.
  Other council fires were here before ours : a classic Native American creation story as retold by a Seneca elder and her granddaughter : the Medicine Stone speaks from the past to our future / Jamie Sams and Twylah Nitsch. — 1st ed.
       p.   cm.
    ISBN 0-06-250763-X
    1. Seneca Indians—Legends.   2. Indians of North America—New York (State)—Legends.   I. Nitsch, Twylah Hurd. II. Title.
  E99.S3S16   1991
  398.2'0899.75—dc20   90-55307
                                                               CIP

91 92 93 94 95 MART 10 9 8 7 6 5 4 3 2 1

This edition is printed on acid-free paper that meets the American National Standards Institute Z39.48 Standard.

A SENECA PRAISE

Oh Great Mystery, we awake
To another sun
Grateful for the gifts bestowed
Granted one by one—
Grateful for the greatest gift,
The precious breath of life;
Grateful for abilities
That guide us day and night.

As we walk our chosen paths
Of lessons we must learn—
Spiritual peace and happiness
Rewards of life we earn.
Thank you for your Spiritual Strength
And for our thoughts to praise;
Thank you for your Infinite Love
That guides us through these days.

Dedication

We dedicate this work to the loving memory of
Grandpa Shongo. We are grateful for all the wisdom
he shared with *The People* and for the lessons he is
still teaching us from the Spirit World.

CONTENTS

Acknowledgments

We would like to acknowledge the staff of Harper-
SanFrancisco for their tireless efforts on our behalf.
In bringing forth the wisdom of the Ancestors, they
have fulfilled a dream. Our special thanks go to
Barbara Dreamweaver for her belief in the project
and her talent in assisting the birth of our history
into written form.

Jamie and Twylah

INTRODUCTION

COUNCIL FIRES
OF THE ANCIENTS

In ancient times, the main purpose of nightly Council Fires was to learn how to listen. The truths of how to live in harmony were kept alive by wise Storytellers who would relate Tribal wisdom through Medicine Stories as those who gathered to listen sat around the nightly fires. Tribal Tradition, history, acts of courage, and lessons on how to discover the true Self came to life through the events related in the legends of the Ancestors. It was the responsibility of the listeners to relate and apply those truths to their personal lives in a manner that would make them grow.

As the authors of this book, we are Storytellers. The responsibility of sharing the voices of our Teachers has fallen to us. Our Teachers have been our own Ancestors, our Totems, the Spirits of the Wind, and All Our Relations in the Planetary Family. This particular story is the history of the Creation of our Planetary Mother and her children as it was passed down through Native Elders and as it was experienced firsthand by the seven-sided Medicine Stone, Geeh Yuk (Seven Talents). The content of Seven Talents' story may startle some people who do not know the history of the Earth from one of the Native viewpoints.

The history of Creation and of the Four Worlds that our planet has passed through has been an oral tradition among our people. The responsibility of maintaining the purity of that history and those legends falls to the Story-tellers who earn the right to pass the wisdom to the next generation. The Prophecies of the Fifth, Sixth, and Seventh Worlds have been accumulated by the Native Seers through the ages and passed to the two of us. Each vision has been shared in our Medicine Lodges, and with those who have the ability to nurture the dreams of the future. Our Medicine People take note of the importance of each Earth change in order to prepare the coming generations for the implied future challenges.

The Allies and Totems of nature have been great teachers to Native Americans throughout time. The Stone Tribe, being the eldest of all who have lived on our Earth Mother, are the Record Keepers. Through the vast libraries contained in the rock bodies of these Medicine Helpers, Native American Storytellers and Seers have come to know a great deal about our history and our future.

The Stone Nation has long reminded our people of the validity of all Tribes of Earth. All five races of human-kind, the Creature-beings, the Plant Tribe, the Spirits of the Sky Nation, the Creepy-crawlers, and the Finned-ones that swim the seas comprise All Our Relations. The nonhuman members of the Planetary Family are not usually seen by other races or creeds as being equal life-forms and, there-fore, our Teachers; however, in our family Grandpa Shongo taught the wisdom of letting go of our spiritual arrogance in order to understand the languages of all other life-forms.

Grandpa Moses Shongo, a Seneca Medicine Man, reminded the human world that "other Council Fires were here before ours." Grandfather Shongo told the stories that kept the history of *The Faithful* and the Red Race alive in the hearts of those who listened. His quote, "Other Council Fires Were Here Before Ours," is carved above the rotunda of the Erie County Historical Society and Museum in Buffalo, New York. This quote is a reminder that we Two-leggeds were preceded by our Ancestors, and they were preceded by the Creature-beings and Plant People, who were preceded by the Stone Tribe. In every generation, there have been those who believed that the wisdom of the Ancients was outdated, forgetting that without the proper use of that wisdom, the present generation might not have survived.

During his life, Grandpa Shongo taught Twylah to honor the lessons of the Medicine Helpers, and he has continued, in spirit, to teach both of us since his death. Geeh Yuk, Seven Talents, was Twylah's playmate and taught her the Language of the Stones, whose markings were over many hundreds of thousands of years in the making. The gift we want to share with the world is the history and prophecy of our world as seen through the eyes of our Ancestors and one of the Medicine Stones of the Earth, Seven Talents.

The symbols that represent the Language of the Medicine Stones are recorded at the back of this book so that each Two-legged human can find his or her special Teaching Stone and understand what the talents and lessons of that Rock Person mean to them personally. The

Language of the Stones can be an oracle, a teacher, a guiding force, a legend of one's destiny, and/or a personal friend. The Teaching Stone's symbols represent concepts that individuals must apply to themselves by using their own intuitive abilities. Personal growth and applying the understanding of every concept is therefore left to the heart's understanding and the beholder.

The faces and forms that can also be seen on the surface of a Medicine Stone contain messages from Allies of the Spirit World. The faces of the Ancestors, the forms of the Totems, and the teachings each has to impart are mysteries that require contemplation and intuition in order to be understood. Each Medicine Stone carries the records of all that has come to pass in the life of our Mother Planet and can assist each seeker in finding his or her place in the great scheme of life. We, as Storytellers, know that each part of nature carries a language that can teach us to further understand the validity of our planetary counterparts. These languages of nature are a way for us to join in harmony and walk our paths of self-discovery with All Our Relations.

One of the most Sacred Ceremonies in our Tradition is the Give-away Ceremony. We have learned that we must share the wisdom, as well as the future prophecy, in order for the Teachings to live. We trust that the story of Seven Talents will touch your hearts and that the Medicine freely offered by the Stone Tribe will assist us all in fulfilling the prophecy of life abundant for the next seven generations.

PROLOGUE

Nyahweh Scanoh, greetings, Children of Earth. My ancient eyes have long beheld our Earth Mother, and many long winters have been carried on my back. Since before the dawn of Creation I have traveled through the Void discovering and rediscovering the beauty of the Great Mystery. Wise I have become, and yet each new change in the Planetary Family continues to allow my understanding to grow further, as if I were a newborn babe. Before the creation of our physical world I was innocent and yet hungry to know the mysteries of life. I have recorded each change completed and lesson learned along the Path of Beauty in order to mark the growth of all life-forms.

I have soared with Eagle high above the canyons of time observing the evolution of the Planetary Family. You and I are alike in many ways. Though you are human and I am a Stone Person, we both feel through our hearts. We both learn through experiences shared with All Our Relations. Every relation has personal missions to complete so that life abundant may flourish and grow. I have been a teacher and friend to many Two-legged humans, teaching them the similarities among all members of the Planetary Family.

I am of the oldest of Earth Tribes, The Stone People. Our Earth-forged bodies have been fashioned from the erupting heat of volcanic creation and the ice-blue cold that tempers our ancient spirits. *The remembering* of each tempered spirit, set in Stone, allows us to hold the memories and records of all that has ever occurred in this world. From other Stone People, who have fallen from the Great Star Nation, we have learned much of other galaxies. The Meteorites who have joined us, from the stars, have added to our knowledge and to the libraries carried within our rock bodies. We have many stories to tell the Children of Earth about the worlds that came before the written histories of the Two-legged humans.

I have waited until now, the dawning of this Fifth World of Peace, to set these stories in motion. I am called Geeh Yuk, or Seven Talents, in one of your Native American languages. I have come to teach you how the Language of the Stones came to be and to bring forth these ancient *rememberings*.

Those of you who have ears to hear and eyes to see and hearts to understand know that these truths are for all races, all people. The Great Mystery created the Children of Earth in different colors so that one day they would come together and compose the Whirling Rainbow Tribe of Peace. Now is the dawning of that special time. The promise of world peace is unfolding. We, of the Stone Tribe, trust that the Path of Beauty will be made clear to all Two-leggeds once they remember the victories and the failures of the other worlds that flourished and then passed into oblivion. Those old worlds have served the Children of Earth by leaving legacies that gently point the way to the new world's prophecy of peace. We, your ancient Brothers and Sisters of the Stone Tribe, are here to remind you that the time is now and that the power of this prophecy lives in you.

CHAPTER ONE

CREATION

Oh Great Mystery, what is this I feel? The vastness, the total darkness, and this sense of well-being. My heart answered my question with a flutter because there was nothingness surrounding me. My mind raced with possibilities, only to be surprised that I had the ability to think. Oh Grandfather, Grandmother, is this a joke you are playing on me? This darkness is limitless and yet I have no fear. "Great One," my mind called out, "where am I?" My senses are reaching out into this eternal Void and feeling the textures of joys I have never known. "Oh Great Mystery," I whispered, "inside of that I am feeling there is a growing presence that baffles me. What are these new sensations?"

My mind sought to identify the emerging phantom of exquisite beauty only to have it merge with me. Shudders of complete happiness overtook me when I melded with the phantom and suddenly I knew. I understood the mystery of the Void. In unparalleled ecstasy, my mind soundlessly cried out in gratitude to the Original Source. "Great Mystery, you live in me as in all other thoughts here in the emerging Void. I give you thanks and praise for this joy I feel, knowing the wholeness of potential Creation and knowing that I belong to it."

My thoughts rolled through the deep nothingness of the Void, touching other ideas and feelings, then came to rest somewhere in the inky darkness. I felt an eruption in the blackness and watched as a brightness was birthed. Light came into being and gave sight to all the inhabitants of the Void. A cloudlike substance began to arise and encircle Eternal Land as Great Mystery created the Field of Plenty. All that would ever be needed in Creation began to form as patterns of light and thought. I marveled as each idea exploded into wondrous expressions of feelings, colors, and tones. The limitlessness of Great Mystery's abundance sent shimmering rainbows of light through me. I felt the goodness and beauty of every new part of Creation as each emerged. Patterns of color traveled through the mists, uniting and dancing with one another in endless space.

As I responded to the ecstatic waves that emanated from every thought, I noticed that I, too, was a patterned set of colors dancing in the light. I reveled in the discovery and observed the changes in me when other designs touched my being. I felt each feeling as music, color, heights of knowing, and joy. Soundless though they were, each thought-form became rhythms of Great Mystery's heartbeat, silently dancing on. Each life-form was a thought of interlaced beauty playing upon every other harmony nestled in this endless space. The Void was full of wondrous gifts of beauty, each evolving and exploding into new heights of love.

fell
↓
listen

Great Mystery noticed that the Music of the Spheres was felt but was not heard. The gift of listening was added to Creation and *Jud-gown* (Eternal Land) was complete. "Oh, Swenio," my heart cried out, "You who are the Great Mystery have given us yet another gift of love that makes my spirit soar." My mind was filled with the presence of melodious rhythms that commanded my feelings until each emotion was felt and explored in fullness. I began to understand that I was everywhere and nowhere, empty and full, all and nothing at the same time. Like me, Eternal Land was an indivisible Uni-world created in wholeness as a unified expression of Swenio's love. A translucent sphere began to encapsulate Eternal Land as the Field of Plenty prepared each unique life-form for manifestation in the First World, which would be called *Da-was-sa-wah* (in the beginning).

Many patterns and colors of light gathered together as Great Mystery rolled some matter from the Field of Plenty into a bubble. The gifts of beauty that would live within the bubble of matter sprang forward and disappeared within the ball. Swenio breathed upon the sphere, giving it physicality, and called it Grandfather Sun. This was the first divine act of Creation that manifested thought into form. Love was the essence of Grandfather Sun. All of Eternal Land was awed by the intensity of his golden light. Great Mystery decreed that Sun would feed all who came into his circle with love. Grandfather Sun's mission was to give warmth, light, and love to all Creation unconditionally. He was to be a constant reminder to all of Swenio's children that each of them carried a part of the Eternal

Flame of Great Mystery's love and that none were excluded from that source of light.

Patterns of silver and blue swirled by my line of vision, weaving a new wonder to behold. The Music of the Spheres softened, and chimes tinkled in delicate harmonies as Great Mystery formed the body for these gifts of beauty to enter. Another sphere, smaller than Grandfather Sun, was rolled from the Void into the First World. The silvery light laced with the palest of blue gently gathered Grandfather Sun's light, taking it into her body and reflecting it outward once again. Grandmother Moon had been birthed from the limitlessness of the Spirit World in a glory all her own. Her gentle beauty touched my heart as she spoke to us of her mission, which was to give light in the places of darkness that would be called night. This second manifestation, Moon, was the reflector of Grandfather Sun's love and the weaver of our feelings, making them into dreams as we rested.

To give our feelings substance, Swenio, the Great Mystery, called forth shimmering waters from the Field of Plenty and gave Grandmother Moon the mission of weaving the ebb and flow of the countless waves of emotion. The dancing, fluid substance called Water encircled Grandmother Moon and joyously exploded into shining droplets coming together again in a body of liquid silver-blue spiraling in the reflected light of Grandfather Sun. Water's songs were varied and spontaneous. As Water touched the other light patterns and Gifts of Beauty, the melodies and harmonies changed because the liquid would reflect the feelings of those whom it touched.

Great Mystery smiled and gave Water its mission of cleansing, refreshing, and reflecting new feeling within all it would touch on its journey. Water was told that it would have two parts to its nature. The refreshing and feeding aspect of Water would be fresh, and its cleansing and releasing aspect would contain salt. The part of Water that had salt in it carried the feelings of sadness away, finding a new and revitalized use for the transformed droplets. The fresh and salt aspects of Water would return to skies of Eternal Land to share the new understanding of all feelings Water encountered with the Clouds and the Sky People.

Great Mystery realized that Water did not fully understand how this process would work, and so the fourth act of Creation began. We all looked deep into the Void as Swenio once again pulled a misty substance from the Field of Plenty and rolled it into a ball that began to manifest as a swirling blue-green orb of light as it took form. Earth was her name, and she was a glory to behold as she sat between Grandfather Sun and Grandmother Moon. Great Mystery spoke to our hearts and decreed that Earth would be the Mother of countless life-forms that would manifest physically when the time came. Swenio then called forth all of the colored light patterns that would be a part of the Earth Mother's body. These future manifestations would live as thoughts inside her form until the time of their births into physicality.

I felt deep within the core of my being a pull so strong that I cried out in joy. I had been chosen. I was to be a part of the Earth Mother. I was to know my talents and missions. I marveled in wonder as I began a spiral dance and flew across the Void spinning in exquisite patterns of frolicking light. New colors and feelings erupted from my core and sent their patterns dashing across the indigo vastness of space.

Just as I was about to enter the form of the Mother Earth, Great Mystery spoke to my total being and etched eternal wisdom and new knowing into my memory. Swenio's voice spoke in the midst of my revelry: "You are a Child of Earth, no less than any of her other creations. You will manifest as a part of her body and will hold the history of her passage in your heart. As will all of my other creations, you will carry my love deep within you, in the form of the Eternal Flame. You will be called Geeh Yuk, which means Seven Talents. Your talents will be used to enlighten other Children of Earth so they may recognize their personal talents and abilities. You are a Stone Person and a teacher for those who seek to know the truth of the Earth Mother's wisdom," *Da Naho* (it has been decreed).

I felt a new emotion coming up from deep inside of me as I held Great Mystery's words in my memory. This new feeling was one of belonging to a whole that honored my mission and my part in Creation. I felt the perfection of Swenio's plan emerge in my new Knowing Systems. Then I was told about my personal talents. "Geeh Yuk," Great Mystery whispered, "your talents will find many ways to express themselves on the Earthwalk you are about to take. Use these talents and develop the abilities that are inside your spirit. These talents are faith, love, intuition, will, creativity, magnetism, and healing. You will know how to find and develop these gifts as you live in the physical world."

When Great Mystery spoke these last words to me I was pulled into the body of the Earth Mother and felt a warm glow surround me with her love. From inside her body I continued to watch the dawning of Creation as Swenio continued to arrange patterns of eternal beauty and set them in motion, dancing and playing throughout the Void. Soon, all the patterns had been placed and given their decrees and missions. Life abundant had quickened and the Gifts of Beauty with their miracles of wisdom began to spiral around with such a speed that their interaction created the harmony that added to the melody of Swenio's creative orchestration of natural order among the orbiting bodies of Sun, Moon, and Earth.

The song of each individual creation melded and became one with all others. The Celestial World filled with the Music of the Spheres, becoming whole and yet unmanifested physically, inside the heart of Great Mystery. This Celestial Spirit World was endowed with wisdom and perfection where life, unity, and equality would be held in sacred trust forever. No discord could penetrate the melodious feelings of at-one-ment that emanated from every spirit as the crescendo of unified hearts came together as one.

Swenio breathed the Eternal Flame of love into the center of each life-form and told us that this inner space would be called our Vibral Cores. That special place inside our beings would be the home of our wisdom, our sense of balance, our understanding of eternal at-one-ment. Upon touching my Vibral Core, I began to develop my first gift of faith. At that moment, I understood wholeness and would trust that feeling forever. The gift of sacred trust would empower my faith and inflame my creativity as I moved forward, spiraling into space, with the evolving whole.

Great Mystery collected substance and added it to the creations of the Spirit World in Eternal Land. As each creation received matter, each moved through the Void and was born in the space of the physical world. Each part of Creation sang gratitude to Great Mystery for the inheritances of beauty, wisdom, love, equality, and wholeness as life was breathed into form. Seeds of growth were placed in perfect order in the vastness of our new physical world. Each life-form was given the ability to give and receive. All was in balance and harmony as life abundant flowed through the beginnings of the new Sky World, which Grandfather Sun, Grandmother Moon, and Mother Earth would now call home.

CHAPTER TWO

THE SPIRIT WORLD

Before the dawn of physical manifestation, I was among the first Stone People to hold Council in the Spirit World. During that Council, the Sky People, who waited in the Void with all others who would be going to the new world being created, told me of the coming of other Tribes who would share the planet with the members of our Rock Clan. Our Tribe would await the appearance of our new Sisters and Brothers: the Plant People, the Creature-beings, the Little People, the Creepy-crawlers, the Water Spirits, the Cloud People, and the Two-leggeds of humankind. In my Seven Talents, I found the potential to dream the future beyond the Void. My dream was a reflection of the story I had heard from one of the Sky People.

In my dream I saw myself speaking to the other members of my Clan. We were perched atop a bubbling lava pool, looking deeply into the fiery eyes of potential creation that spurted from the Earth Mother's body. The scene looked very strange to me since I had never seen any of the Spirit World ideas as they would appear in manifested form. I was surprised to see that my rock body had seven sides and was a dusty color of brown. I went further into my vision in order to hear the words I was speaking to the other Stone People who had gathered. It was then that I heard the ring of my own voice for the first time.

"The time is coming when we will share our Earth Mother with other children to whom she will give birth. There will be four new types of people manifesting among the Stone Tribe, who are the only Children of Earth who have taken form. These new Sisters and Brothers will share our Councils and the love of Grandfather Sun. The Sky People, who have existed in Eternal Land as spiritual beings since before the dawn of Creation, will add some of their numbers to those who are coming to join us. It was one of these spirits who told me of the wondrous expansion that is about to add different varieties of life-forms to the Planetary Family.

"This Sky Person spoke of their Tribe," I whispered. "My heart was gladdened to hear of their mission among us, for they have minds of great wisdom and limitless understanding. These spirits of the Sky have beautiful faces but no hearts or bodies, as they are not regulated or aged by time. Their mission is to work as Thunderers under the direction of Hinoh, the Thunder Chief. They will make their presence known to us as they send Water Spirits to the body of the Earth Mother; they will create oceans, rivers, streams, and brooks to nurture the seeds of others who will follow. These Sky People will show us their faces as Grandfather Sun returns their water essence to the Sky Nation in the form of Cloud People," I continued.

I smiled as I told of the extra gift, a miracle of the Sky People's mission, to the others of the Stone Tribe. "Great Mystery has also given these Sisters and Brothers another option in their mission. If they choose to experience physical limitation, they will be allowed to take the gifts of wisdom they hold and enter the Earth, taking hearts and bodies in order to experience the feelings of the heart through touching and sharing." This last revelation brought a whir of whispers as each Stone Person reflected on the beauty and simplicity of Great Mystery's Divine Plan.

In my dream, I waited for silence before I continued, "The Sky People contain the balance of ideas in their natural forms. They supply these ideas to the Children of Earth as all Tribes seek the harmony of the Earth Mother's Circle. Each Earth Tribe will seek wholeness in its individual way, leading each individual to truth. When the level of self-understanding is reached, the Children of Earth will become whole entities. The Sky People will assist all Tribes in uniting an earth body and their mental gifts. Every time a spirit walks the Earth, a new circle of lessons will be learned in a new body with new understandings."

Smiling at the awed expressions of those gathered around the Council Fire, I spoke of the other people who would join the Sky People. "The Little People are others who have lived with the Earth Mother since the beginning of time. They are the Tiny-ones who will look very much like the Earth Tribe of Two-leggeds. The Little People will live in the hollows of old tree stumps, in tiny mounds of earth near streams, and deep in the green forests. Some

of these smallest of people will wear plant leaves for camouflage or covering for their bodies. Others will be formed like the Two-leggeds but will also have wings, and others will appear as shining lights in the dense forest.

"There are three kinds of work among the Little People. Some are hunters, finding lost objects. Some are stone throwers, assisting the Stone People in changing location. The others look after the seeds. In the miniature world of the Tiny-ones, they wake up the Plant People and paint their fruit when it is ripe. They can also assist the Human Tribe in a variety of ways when a need arises.

"The Creature People are a vast and varied Tribe and will make up what will be called the animal kingdom," I said, making my voice encompass the grandeur of a gesture I literally could not show because I had no arms. "They come in many colors, forms, sizes, and shapes. These Creature-beings will exemplify the instincts needed by the Earth Tribe in their new human forms. Some of the Creature-beings will fly, some will crawl, others will walk on four legs, and some will swim the waters brought by the Cloud People of the Sky Nation. The wise teachers of the Creature-beings know how to live in balance with the Earth Mother, maintaining harmony in Nature Land. These Tribal Members understand how to hunt and how to be hunted. Through their survival instincts, they will teach many lessons to the other Children of Earth.

"The last group, who will be called the Earth Tribe, are human Two-leggeds." Even as I explained the humans to my fellow Stone People, I could not imagine them. In my dream I saw only what I was saying, not how it would actually be. I was concerned and curious as I listened next to my own words, wondering how the seeds of future would come into being.

"There will be five races and colors of these Two-legged Earth People. These human Children of Earth have bodies that stand upright in order to support their minds, keeping their thoughts above their bodies as a reminder that thoughts are the first act of manifesting reality. These gifted Two-leggeds must seek wholeness by balancing their human needs and their intellectual understanding. The Earth Tribe People will increase their knowledge and wisdom through reaching out to touch all other living creations in our world. In seeking answers, they will need to use their natural potential and then share those gifts with others."

I wondered to myself what these funny creatures would look like. I had seen the multitude of colors that composed each spirit in the Void and wondered which five of those glorious hues would mark the five races of Two-leggeds. Were they going to be blue or orange? Would they have long appendages to support their enormous minds and intellects above the round, lump bodies similar to the bodies my Stone Tribe seemed to have? I was puzzled by my dream's connotations, but having had no experience of the physical future, I could only continue to muse and mull over the possibilities. I decided that Great Mystery must have a better imagination than I did, so I went back to my dream and began once again to listen to the words I was speaking in that Dreamtime reality. Maybe I could see something in my dream that would help me figure out what each of the strange new life-forms coming to Earth would be like.

"All these new Children of Earth will be welcome at our Council Fires," I proclaimed with a certainty that made me feel foolish. There I was in the future, talking about something about which I still had no clue. "Each Tribe carries many gifts we can record for *the remembering.* The Stone Tribe will hold the wisdom of each so that the Eternal Flame will glow brightly for the future generations. We have waited for eons for the coming of these Sisters and Brothers. The Sky People hold the wisdom of eternal limitlessness, the supply of intelligence, new ideas, and the mind housed in spirit and body. The Little People supply the space and place for change as their miniature world affects the larger world as a whole. The Creature People carry the desire for maintaining a natural environmental identity of harmony with the Earth Mother and life with her. The Earth People provide the vehicle for attaining the goal of wholeness. Together, the circle of all these Tribes will be the symbol of harmony created by Great Mystery. As we hold Council, listening to one another's needs, there will be peace throughout the land. We will always come together in a circle to honor the harmony and wholeness symbolized by Great Mystery's Divine Eternal Plan."

As my dream faded from my mind, I took great pride in the wisdom I would seemingly gather from future experience in the manifested world. It was hard to imagine, but it was obvious to me that my future was going to be bright, because all of the Tribes who would inhabit Earth would work hard to live together in peace. At that moment, I decided that life on Earth would surely be an adventure in discovery. I was expectant and happy about the possibilities of seeing these beautiful lights in the Spirit World manifest in their various forms. I never worried that their differences would create any problems. No one in the Spirit World had ever heard of jealousy or greed, so in our innocence, we set out to discover the joys and pleasures of the coming manifestation.

CHAPTER THREE

THE FIRST WORLD OF LOVE

The First World manifested in its splendor and glory as Grandfather Sun beamed love on all of Great Mystery's creations. Grandfather Sun took his place in the Sky World and Mother Earth began to move in her own orbit around his shining light. Grandmother Moon found her circular path around the Earth Mother's orb and began to weave the reflected love of Grandfather Sun into the darkness called night. The orbits of these celestial bodies set and regulated cycles that would mark the passage of Time in the physical world. Time was a new concept, for all ideas of change were constantly present in the Spirit World. The Spirit World was also called Eternal Land because of the infinite, spontaneous, limitless creation that happens all at once and has no beginning or ending.

When the spirit of the Earth Mother had descended from infinite space, we, the Stone People, were the first to manifest. Destined to be the recorders of truth, who would compose the body of Earth, we moved forward and took our places as symbolic stepping-stones that would lead humankind and all other creatures to understanding the

Foundation of Truth from which all things would grow. Water had not been called forth from the Spirit World and waited patiently until the moment when its life-giving forces would be needed by those of the Stone Tribe who had contributed part of their bodies in order to produce soil.

I felt the presence of others of my kind as we bound together to provide the solid base for the Earth Mother's body. Heat from her molten heart spurted in brilliant orange and red hues, giving us the colors of kinship, faith, and passion for life, which bind the Stone Tribe together as one. The Foundation of Truth was based in love and allowed the mountains to form, giving feminine grace to the Mother's body. Her sweet, gentle voice echoed within the hearts of the Stone Tribe as we felt her nurturing encouragement, asking us to form and re-form the boundaries of our common foundation.

We bound together, sharing our gifts and exploring every avenue of our common bond of love, time and again. With each new bonding we found new abilities, which manifested as soil, volcanic ash and lava, hills, mountains, valleys, and layers of solid knowing. The Stone Tribe created vast libraries of rock, each contributing to the whole as the Earth Mother gave thanks with approval. At each joyous explosion of raw creativity, the volcanoes erupted, bringing forth new liquid rock bodies forging our love of truth together as one. We felt the wisdom and the caring the Earth Mother had for our Tribe as each new eruption brought her thanksgiving for our joining together with her. As the first and oldest Tribe of Earth, we rejoiced in the creation of our Clan. We celebrated, giving words of praise, to honor the uniqueness of each intricate rock formation, the beautiful colors, shapes, and patterns of each Mineral and Stone in our family.

Some Stone People would be used for healing, others for fuel, others for their mineral contributions, some for protection, others as brilliant colored gems; but all would be used as libraries of Earth records that imparted wisdom as well as beauty. Every member of our Earth Tribe was full of talents and abilities. We were tempered by volcanic heat and would be cooled when Water was called from the Field of Plenty to fill the valleys we had created as receptacles for the Earth Mother's new liquid Tribe, the Water Spirits.

The Earth Mother spoke to us through our hearts, telling us of this new Tribe, filled with liquid essence, who would cool her surface and bring the life-giving Rain People who would allow all things to grow upon her surface. She spoke of the other spirit patterns that Great Mystery had placed in her care. These would manifest as the seeds of all Plant-beings, which would grow into trees, fruits, herbs, flowers, and vegetables to feed the Children of Earth throughout all time. Great Mystery had decreed that the seeds would bring forth life when nurtured in soil and fed by the liquid life-force of the Water Spirits. The mountains had grown deep and had sloped wide to hold the fluid bodies of the Water Tribe. The Stone Tribe was ready, and so it was that the Tribe of Rock People cried out in welcome when the first Rain People began to arrive.

The Water Spirits came and filled in the vacant spaces we had created for them. All over the Earth there were running brooks, wide rivers, lakes, streams, ponds, and creeks creating steamy pockets of water that touched the soil rich in growth potential. The Water Spirits returned to the Sky Nation time and again, creating Cloud People who blessed the Mother with constant, gentle rains until her surface had cooled. The Stone Tribe watched in awe as the process was continued and the passing moons turned to many winters and the winters turned to hundreds of rotations around Grandfather Sun's brilliant form. Eons had passed before the time was ripe for seed Sisters and Brothers to appear.

The Water Spirits, who had existed in the Spirit World before they came to Earth, had now become Water-beings as they took form as raindrops in the sky. These new Water-beings then fell to Earth to join their Tribe in becoming streams, rivers, and finally seas. They fulfilled their mission of fertilizing the Earth's virgin form, and soon life abundant began to spring forth from every pocket of rich, fertilized soil. Every type of Plant Person who would ever inhabit the Earth was blooming, producing, and giving of its fruit. The new Water-beings began to venture out of their original homes, and they created seas and oceans to receive the growing bodies of liquid. The Water Nursery of Creation was born, and the Earth Mother smiled as Creature-beings began to crawl from the seas onto the land. This exodus was the beginning of a new Earth Tribe. The Creature Tribe had begun in the watery womb of the Earth Mother and would continue to develop for many eons as life quickened on every part of the planet.

Every movement and growth spurt among the Plant Clan and Creature People was recorded by the Stone Tribe as we marveled at each new discovery. The small Creature-beings who had crawled from the warm Water Nursery of Creation were cold and huddled together on the Earth Mother's soil. They were forced to explore the land, which was so alien to them, in order to find food and warm lodging places. Great Mystery sent the Field of Plenty to touch the body of the Earth Mother in order to help these newborns. The cloudlike substance touched the surface of the planet, and all Creature-beings heard an inner voice suggesting that they eat the seeds being released from the Field of Plenty.

I noticed that the bellies of these newborns began to expand and were filled. They began to act in a curious manner as they stopped huddling and danced without rhythm as if to shake off the cold. Then each creature in its own way began to shudder and leap, bounce or roll, gallop or flap its arms. I was sitting among other Stone People who excitedly watched with me as a miracle appeared before our eyes. These Creature counterparts were sprouting feathers to warm their bodies and wings to give them the ability to fly to their warm lodges in the trees. The Standing People of the Tree Tribe were happy for the company and welcomed the new arrivals.

The Winged-ones were the first of the Creature-beings to venture out into the First World away from the water. The inner voice from the Field of Plenty, which continued to instruct the creatures, told the Winged-ones that they would be given a physical characteristic that would remind all Tribes of their story. The feet of all winged creatures would always carry scales like those of the Finned-ones of the seas to remind all of the original ocean home of those who would now soar high above the Earth.

I heard the voice as it was speaking to my new Sisters and Brothers and watched their reactions as they marveled at the clawed talons and scaled legs that supported their fully feathered bodies. Some of the Winged-ones began to preen their new feathers and flap their wings, while others played amidst the waves or pounced upon the shore making talon prints in the soft sand. I marveled as each began to make sounds that reminded me of the music I had once heard in the Spirit World. Together, all the species of the Bird Tribe melded their calls into a joyous sound that filled the air and was taken by the Wind Spirits to cover the Mother Planet.

Upon hearing the song of success sent by the Winged-ones, other Creature-beings began to emerge from the Water Nursery of Creation. These other animal species took heart, trusting their own inner knowing and the support from the Field of Plenty, seeing by example how their needs would be met. The Finned-ones who would no longer be sea-dwellers walked the Earth as Lizard, Crocodile, Alligator, Iguana, Serpent, and other reptiles. Being cold-blooded, these reptilian relations would constantly seek the warmth of Grandfather Sun and would remind the Planetary Family that light could warm the coldest of hearts.

The mammals of the oceans, who would later be at home in water and on land, then emerged to begin their transition process. These Four-leggeds, who would wander the mountains and valleys of the inland heights and plains, came forth. Upon eating the seeds of life, they began trans-forming by growing fur and stable, sturdy legs. Every Four-legged was given a way to develop survival instincts that would later serve others who would seek that wisdom.

Next, the Little People emerged from the frothy waves, having come from their hiding places among the giant shells and coral reefs. They gladly took their missions as keepers of the seeds, stone throwers, and hunters. If a Stone Person needed to be moved to another location, we could call upon this Tribe of Tiny-ones to assist us. They would keep the seeds of all earthly plants in good order and readily available for all time so that no type of plant would ever become extinct. If anything on or below the Earth Mother's surface was lost, the Little People would be in charge of finding it.

I watched excitedly as the Wee-ones began to gather the seeds that had been deposited by the Field of Plenty. Their chattering and squeals of delight carried far and wide to bring joy to the other newborns of our growing family. The Little People were quite a creation. Each was uniquely molded from a variety of shapes and forms. Some were slender and others were stout, some had wings and others had two legs, which were hairy like those of the Four-leggeds. Never in my wildest dreams could I have imagined the myriad of physical shapes that now danced before me. I was surprised and overjoyed to see the way in which they took to their work as they collected and arranged the seeds that would bring green and growing life to the Earth Mother. The laughter of this tiny Tribe was contagious, setting a new type of expression in motion among the newborns and Stone People who watched. Mirth was born, and our gladness rang through the Field of Plenty as Mother Earth joined us in the celebration of life.

Finally, the Two-leggeds came from the Water Nursery of Creation and stood on the shore. These creatures were to be called human and would be no greater and no lesser than any of Great Mystery's other Tribes or Clans. There were Earth People and Sky People among the Two-legged Tribe. The Sky People had come from pure intellect and the omnipotent mind of Great Mystery. Some had chosen to leave the Tribe of Thunder-beings, Cloud People, Stars, and Moons in order to experience the limitations of being physical. These Sky People had chosen to receive hearts and bodies in order to feel, love, and learn through the limitation of being human. The Earth People of the Two-legged Tribe were no less courageous in their missions, because they were to learn the lessons of harmony with all other life-forms on the planet. These beautiful Earth and Sky People of the Two-legged Clan were Guardians of the Sacred Hoop and the Planetary Family. It was their mission to keep the heart fire of the Eternal Flame burning brightly throughout all seven worlds of time. The changes that would occur during those eons would shape the destiny of Mother Earth and all her children.

I marveled at the shapes and five colors of the humans. I inwardly blushed as I remembered how I had imagined them when I awaited manifestation in the Spirit World. I noticed that my innocence had lacked experience and my imagination had been incomplete before manifestation. I struggled with my former images of Two-leggeds. These humans in front of me were neither orange nor blue. They had two legs and two arms, which were formed to balance their torsos. They had fur of different textures and colors just like the Four-leggeds, but they did not have as much of it. They had eyes that reflected the light of Grandfather Sun and shone in different colors. They were female and male just like the other members of the Planetary Family, but my former imagination of them had never included the perfection of form that Great Mystery had produced. I was deeply humbled as I reviewed the wonderful forms of all of the Tribes and Clans that had emerged from the Water Nursery of Creation. I watched in wonder as all of Earth's Children, animal, plant, human, and stone, touched one another without fear and began their discovery process.

Instinctive harmonious living was present among all of these new Tribes of Earth as each set about the task of learning the lessons of living together and in balance on the Earth Mother. All parts of Creation understood their roles in Great Mystery's Divine Plan. They lived according to a plan of survival that allowed the strongest to provide the seed for the next generation. Each species kept its bloodline pure and distinct. An inner knowing identified and defined the natural way of life for every member of the Planetary Family. All of Earth's children marked and honored their Sacred Spaces and respected the territorial boundaries of others. The balance of peace had taken root, and the First World sang the praises of Grandfather Sun's warmth and love.

I kept on hearing the song of my own heart as I blended with these new Tribes. "Seven Talents, Seven Talents," my heart sang to me. I knew who I was and what my seven abilities were, but I was unsure of how to use my gifts to grow and help others. I knew that in giving life to others, the Planetary Mother could fulfill her destiny and purpose. I knew that in providing for the needs of all of Earth's Children, the Field of Plenty would fulfill its mission. I knew that the Four-leggeds would teach survival instincts for other Tribes in order to fulfill their destiny. I knew that I would record the truth of Earth's history, but how I would share that wisdom and with whom still escaped me. The Earth Mother looked kindly on me and spoke to my troubled heart.

"Seven Talents," she whispered, "You are a part of my body. You hold my wisdom, and yet you are not using your gift of intuition. Who among my children need your assistance the most?"

I looked within my Vibral Core and there was my answer. "The human Two-legged Tribe needs me, Mother," I replied. "Some of them are weak because they came from the Sky World and did not have hearts until they arrived here."

The Earth Mother smiled at my revelation. "You have found your answer, Geeh Yuk," she beamed. "You also carry the gift of love. It is time for you to share that gift with the Sky People of the Earth Tribe known as Two-leggeds. Remember that this First World is the World of Love and that yellow is the color of that love as seen in the rays of Grandfather Sun's light. Share that knowing, Geeh

Yuk, with those who have never had hearts before. You will find them in all five races of Two-leggeds." Mother Earth sent me on my way to find a Little Person to hunt for the Sky People with brand-new hearts.

When I found Nog he was counting seeds in the knot of an old tree stump. Another stone thrower had tossed me his way, and I landed at Nog's feet with a start. Nog was about two feet tall with white hair, wood-grainy skin, and a contagious smile. Naturally Nog was curious as to why I had suddenly landed at his feet asking for assistance. When I told Nog what I was looking for, he looked a bit puzzled, since he had seen only two colors of Two-leggeds.

I began to explain that there were five races or colors of humans. The black race was the first to appear and had been endowed with the gifts of seeking the truth of the Void of the unknown. This darkest of the five races would set up rituals for seeking inner peace and growth. These black Two-leggeds would use rhythm to dance and connect to the heartbeat of the Earth Mother. These ebony Children of Earth would be able to withstand the strongest of Grandfather Sun's rays of light and would carry that love into the world through the expressions of truth they found in the Void.

I then explained to Nog that the brown Two-leggeds were connected to the records of Earth and would provide the knowledge of survival instincts for all of Earth's Children. These brown humans would work with the soil and respect the unity of working together for a common goal. Then I shared with Nog my understanding of the red race, who were faithful Guardians of the Earth Mother's body and her creatures. The red people would honor the Mother for her selfless giving and would set up the understanding of Sacred Space, respect, fair judgment, and Sacred Points of View. This red race would further the understanding of the Sacred Hoop, the circle of the Planetary Family, which included all creatures who had red blood and all Stone People, Plant People, and Little People as well as the Clan Chiefs of Air, Earth, Water, and Fire.

Then I began to tell Nog about the white and yellow races that he had seen but never spoken with. I explained that the yellow race of humans was here to refine the truth and make new sets of Knowing Systems. These yellow Two-leggeds would teach the others about Ancestors, family ties, and social customs. The other Clan, the white race, was brought forth to use creativity and magnetism to seek the truth. The white-skinned humans were originally gifted with authority, and their mission was to learn how to use those abilities in truth and fairness. When I had finished, Nog grunted his approval and agreed to help me find the Sky People, who needed to learn how to love.

Nog carried me close to his heart, in a pocket made of leaves, visiting Two-leggeds of all races. We found the Sky People and the Earth People among the Human Tribes and taught them all what it was to have a heart, to love, and to be loving. Our constant guide was Grandfather Sun's golden, loving light as we traveled the Earth. The Children of Earth learned a great deal in those beginning times and shared all they had with one another, for the fear of scarcity had not yet touched their hearts.

I had a chance to develop my seven talents as I interacted with these Two-legged humans. Many times I used my gift of faith when I doubted my ability to teach the lessons of love to those who had never felt or loved before. The love I carried inside of me allowed me to stretch beyond my own limitations and to intuit what piece of information would open the human mind to new understanding. I used my talent of creativity every time I came to an impasse and verbally constructed examples that made the concept of love easier to comprehend. Many healings took place when the Two-leggeds realized that they could love more than one person at a time without excluding any member of their Clan or family.

Often, Nog would want to give up the hunt for other members of the Planetary Family with whom we could share our understanding of Grandfather Sun's love. In these times of doubt or exhaustion, I would have to will myself to continue with or without him. I then used my gift of magnetism to excite him about the possibilities of seeing new lands and other customs of the Two-legged people. Nog was learning many new concepts and customs. He wanted to learn more, but he got tired of the constant traveling when we came to the parts of our journey where we didn't run into any Little People with whom he could exchange stories.

Our journey finally ended when we met others who also had been instructing the Human Tribe in the lessons of unconditional love. Raven had seen Hummingbird teaching the lessons of love and Knotty Oak had heard Willow talking to the children of the Two-leggeds as they gathered berries in the glen. Yes, it seemed that the mission of the First World was coming to a close and that Nog and I could return to our home. Just then the sweet Wind of early summer brought the fragrance of Jasmine and the whispering voice of the Earth Mother. "Geeh Yuk, your intuition serves you well. The lessons of love have been shared and the hearts of all of my children are happy. You and Nog have done well. Continue to develop your abilities, Seven Talents, that you may grow in joy. Home is never outside of you; it is where your heart is happy. Feel my love and thanks until we meet again." The breeze that brought the Mother's voice was gone, but our hearts held the memory of her nurturance.

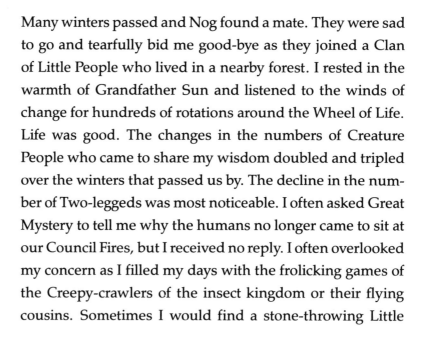

Many winters passed and Nog found a mate. They were sad to go and tearfully bid me good-bye as they joined a Clan of Little People who lived in a nearby forest. I rested in the warmth of Grandfather Sun and listened to the winds of change for hundreds of rotations around the Wheel of Life. Life was good. The changes in the numbers of Creature People who came to share my wisdom doubled and tripled over the winters that passed us by. The decline in the number of Two-leggeds was most noticeable. I often asked Great Mystery to tell me why the humans no longer came to sit at our Council Fires, but I received no reply. I often overlooked my concern as I filled my days with the frolicking games of the Creepy-crawlers of the insect kingdom or their flying cousins. Sometimes I would find a stone-throwing Little

Person to toss me into the brook for a wash and would ride down the stream in the company of Trout. Raccoon would pick me up downstream and return me to my home at the base of Sacred Mountain.

Life was good in those times, and I learned much from the Creature-beings as we shared our wisdom of instinct and survival. The volcanoes and landslides had nearly stopped, and the Earth Mother was coming to a resting point in her process of evolution. It was much more peaceful than in the early part of the First World, when the ground shook almost every day. I reminisced as I recalled the Stompers of the Dinosaur Clan who had pounded the soil every time they took a step. Most Stompers had moved to more tropical regions now. They preferred the jungles closer to the active volcanoes. The earthquakes still rumbled in the areas that needed further expansion and formation, but here, on Sacred Mountain, all was serene.

One day a little girl came to the glen in front of my home and sat amid the Dandelion Clan. She knew the language of the heart, which Nog and I had taught the Two-leggeds. This First World Language was spoken silently with signs and expressions of the face. This Language of Love was called *Hail-lo-way-an*. The child's name was Ka-Nu and she was looking for someone to talk to. Snowy Owl gave me a ride down to Ka-Nu's resting place and we began to share the secrets in our hearts. Ka-Nu told me of the abuse of love that had swept the dwelling places of the Two-leggeds.

The Two-leggeds had worshiped the love and warmth of Grandfather Sun, honoring the golden light of his love, for many generations. One of the yellow humans had begun to convince the other Two-leggeds that the yellow race was superior to the other four colors. Since their skin was yellow like Grandfather Sun, they said, they were loved more than the children with other skin colors. A deep sadness filled my heart upon hearing this news. Ka-Nu continued and told me that a yellow metal had been discovered by the yellow Two-leggeds and that they were hoarding it, saying that it was physical proof of Grandfather Sun's love. The race that obtained and used this metal to make shrines to their Sun-god would find favor and would rule the other races.

I felt a sudden pain in my Vibral Core as the heart within me was torn with grief. Greed had seduced the Two-legged Earth Tribe. My heart cried out to Great Mystery in pain, "Oh, Mystery of life, how can this be? Has all of our work been undone? Great One, please tell me this is not true. I do not doubt this child, but my heart screams no, no, no. Please tell me what is to be done."

Ka-Nu sat in silence with tears etching their way down her soft pink cheeks. She had internally heard my prayer and shared my pain. We sat in the glen, Entering the Silence by stilling our minds and awaited the answer that we trusted would come. All the Creature-beings of the animal kingdom who had felt the ripple of pain sat silently holding council with us.

The Clan Chief of the Wind brought an answer to the hearts of all gathered. The voice of Great Mystery spoke to our hearts. "Children of Earth, the time has come for each of you to learn through opposites. Your strength and growth depends upon your ability to remember and follow the Path of Beauty, which is founded in truth and love. Those who have chosen the crooked trail will have the opportunity to learn what it is like living in a world of scarcity, greed, mistrust, resentment, hatred, and bitterness. Each of you must hold the essence of love in your hearts, honoring the Eternal Flame. Your strength will come from this flaming life-force, which is the part of Me that dwells within each of you. This lesson will be repeated time and again as new spirits come to walk the Earth Mother and partake of her goodness. The road will be difficult and will last throughout the seven worlds of time, but life abundant will flourish again. Find the trust within your Vibral Cores and know the truth of my words."

All who had gathered gave thanks for the message from Swenio, the Great Mystery. We came together in a circle of love and asked for blessings of truth for those who had taken the crooked trail, asking that enlightenment and understanding come to every Sister and Brother. The Earth Mother's voice was heard traveling on the Winds of Change: "Sweet children, you who are faithful to the ideas of unconditional love, you who give gratitude and praise for the blessings of abundant life, know that I will honor your faithfulness and find a way to protect your lives when the time of cleansing comes. Listen to the voice of truth within your hearts. Hear the Language of Love when I speak to you, and follow the instructions I will give. The time is drawing near when you will need to go below the surface of my body in order to survive. You, along with others of The Faithful, will be the beginnings of the combined race called Subterraniums or "Live-Below-The-Earth Peo-ple." I will make preparation by hollowing out tunnels and caverns for your new home. I am your true Mother and you may depend on my protection."

After the Earth Mother's voice subsided, we made a bond of love and promised to share all our personal prophe-cies and visions so that we could travel to the inner Earth together. Many winters passed as we gathered the faithful members of Earth's Children and the seeds we would need to transport. Nog and his mate had returned and brought many children, each carrying a load of seeds that would be planted in our future subterranean home. We were called to the glen when Wind was wailing the Earth Mother's signal to The Faithful. The Thunderers rattled nature's message into the night, awakening the Standing People who would assist us. The Water Spirits sent freezing rain into the previ-ously calm summer night, and we knew that the time had come. Now we must leave the surface of the Earth Mother and travel to the safety below.

The twisted roots of the giant Hemlock uprooted, showing the path to the world below. We helped each of The Faithful through the hole in the Earth and took great care with our precious cargo of seeds. The seeds were our promise of a new world and held the potential of better times to come. The journey to our new home was long and tenuous. We were a curious-looking band traveling along the inner paths of the subterranean tunnels. Every kind of Creature-beings, the Stone People, Creepy-crawlers, and every color of Two-legged was represented. The Winged-ones flew in front of the group and acted as scouts. We met others along the way who had heard the call, responding from all parts of the single land mass of our world, Turtle Island. The destruction of the First World was a cycle of death and destruction but it was also a cleansing rebirth.

When we reached the caverns that would be our new home, a misty light covered the huge open area above. We were told that this was the light of the Earth Mother's inner sun. There were vast areas for growing, and through the mists we were sent gentle, life-giving rains. The water was collected and saved for future use, reminding us that we needed to make some decisions about our new lives in this startlingly different environment. We held Council Fires as we had when the Stone Tribe was the only Tribe on Earth, around a small, bubbling pool of lava. The decisions and plans included the needs of every life-form and produced a common respect that was honored by all. As a group, we recreated a harmonious living situation based on truth and beauty. Life continued in the world below for many generations according to Great Mystery's original plan.

During those times, living in the inner Earth, we were given constant reminders that some would need to choose to return to the Above World when the cleansing was finished. When that time came, I chose to be a part of the new beginning on the surface. I felt that my seven talents were needed to support the beginning of the Second World. Through our Councils we had come to understand the lessons of the First World. One lesson was unconditional love, which was replaced by the fear that Grandfather Sun did not love all his children equally. The sun worshipers had developed the fear of darkness, and from that darkness had come greed and jealousy. Racial hatred and lack of faith had supplanted the true sense of worth that Great Mystery had given every member of Creation before the beginning of time.

The First World had brought forth two kinds of people: *The People* and *Those People*. The People were the members of every race of Two-leggeds who had been faithful to the Great Mystery's original intent of equality and unconditional love. The People honored the lives and Sacred Space of all life-forms. Those People chose to follow the crooked trail of greed and controlling others, ambition, and glory for the self. Those People became a catalyst for all time that kept The People of The Faithful constantly seeking harmony in order to balance the hurtful actions of Those People.

The legacies and lessons that had formed the decline of the First World of Love had to be remembered by The Faithful, or The People. When Those People chose to alienate others among the Planetary Family through dominion over them, Grandfather Sun was called on to purify the pestilence present on the Earth Mother.

The First World was then destroyed by Fire. The Fire Spirits who lived in Grandfather Sun's fiery orb and the Earth Mother's molten core came to burn away the darkness that had eaten away at the Foundations of Truth. When any foundation was no longer solidly built on unconditional love, the shadowy characteristics inside every Two-legged were able to take control. The Fire Spirits had cleansed the world, and now it was time to see the dawning of a cleansed planet. Those among us who had chosen to return to the Above World watched in awe as the Earth Mother and Sky Father brought forth Great Mystery's promise. The Whirling Rainbow of Peace filled the Sky Nation with vibrant colors. Every color was brilliantly displayed, fulfilling the promise of the rebirth of Peace. Our hearts filled with gratitude as we beheld the Aurora Borealis, which came from the Earth Mother's inner sun.

This new rainbow light was the promise that all creatures were loved equally by their true parents, the Great Mystery, Mother Earth, Father Sky, Grandfather Sun, and Grandmother Moon. Mother Earth spoke: "Children of Earth, Great Mystery has decreed that from this moment we are building a Uni-world rather than a Universe. I will always protect and honor those who serve All Our Relations. Allow your hearts to be illuminated with this rainbow light. Find the new rhythm of oneness that will begin the next world, and continue to love one another."

My heart was full as I traveled to the surface with the parts of my family who had decided to return to the surface. The Tribes and Clans who remained below became the race known as Subterraniums or Lives-Below-the-Earth People. The part of our family that had become Lives-Below-the-Earth People bid us good-bye, assuring us that the inner and outer worlds of our planet would always be a bridgelike symbol of all worlds being created in equality, representing unique reflections of the unified world or one world.

So ended the First World.

CHAPTER FOUR

THE SECOND WORLD OF ICE

As the great ice mountains carved crevices across the undulating body of the Earth Mother, the Second World began. The First World's torment and upheaval had crippled the unfaithful Children of Earth, creating fear and a latent understanding of their misdeeds. After the fire of purification had killed most humans, the love of Grandfather Sun could no longer touch or heal them, for the frozen tundra of the Ice World was etched into the face of the Mother Planet. The jealousy and greed of the First World had destroyed the warmth of Grandfather Sun's unconditional love for all creatures and all life was blighted by the cold.

Those who had slipped into the subterranean tunnels, long ago before the rains had stopped coming and Fire had purified the above world, were fearful of returning to the surface. The Little People spoke of the greedy and faithless who had shunned the warnings and omens that had rumbled up from the Earth Mother's belly. I felt the warm rush of compassion fill my body of stone when I remembered those humans I had known in the First World who had died because they had wandered from the truth. I wondered if they would return to new Earthwalks to learn those lessons again? Would I meet them or be their teacher once again? What would the surface world look like now that the purification was complete? I pondered these questions as our band of Ice World settlers neared the opening leading to the outside.

Here, inside the inner Earth, I was safe and warm. I grew in wisdom as I watched the Creature-beings calm the fears of the Two-leggeds who had been born in the subterranean chambers for generations. These Children of the inner Earth had never seen the outside and had no foundations on which to build, except the stories that had been passed down from their Ancestors. Among our band I saw the Little People, who had much longer life cycles than the humans, trying to instill confidence in the trembling Two-leggeds waiting for the signal to reemerge. The Little People realized how their comments about the former purification had upset their human counterparts and began to assure the others of the Earth Mother's promise of a good new world.

Finally we reached the top of the tunnel, entered a cave, and emerged on the surface. Looking upon the face of what had once been a lush, green world, we who remembered gasped in horror. The blue tundra had covered the upper world, leaving only bare patches of green stubble. What would the Children of Earth eat now? The fruits and vegetables were gone. The warmth and shelter of the Standing People had disappeared. Life would be hard for the human Two-leggeds, who had no survival instinct as did their Creature-being counterparts. I was grateful that I would not be a burden to my companions. My rock body did not need to be fed. I had been created whole, self-reliant, and self-sufficient at the dawn of creation. I had to use two of my seven talents, intuition and wisdom, to solve this new challenge of finding food for the others.

It was hard in those first moons for the naked Two-leggeds. I traveled with a young woman named Eka. She rubbed my surface for protection and companionship. Eka understood me and traveled with me in her dreams. I could speak to her of the things she remembered from the First World even though she was not there at the time. The Two-leggeds of the brown race carried the racial memory in the enormous frontal lobes of their brains. Eka was resilient even when her belly was empty, but I hurt for her when her strength grew small and her stomach rumbled in hunger. I called out to the Stone People who made up the mountains and eventually they sent me word, telling us to return to the cave in order to provide shelter for the Two-leggeds. It was clear to the Plant People, Creature People, and Sky People that these Two-leggeds would not survive unless they had some survival instinct. The Sky People sent a lightning stick to a tree stub and Fire came to the children of the Two-leggeds. The cave was warm and every night an ember was saved, stoked, and guarded for the next day's use.

Antelope came to our cave as Grandfather Sun's pale light climbed high in the ice-blue Sky Nation. "Seven Talents," he said, "These Two-leggeds are among The Faithful. They have honored the love of the First World and now they are sickened from lack of food and exposure to the bitter cold. We must hold Council and I must speak to their minds, for they do not understand our sounds or language." I agreed with Antelope, and that night around the Council Fire he gave a great gift to humankind.

The fire was blazing as the humans huddled together in their nakedness as close as was possible to the warmth and tawny light of the Council Fire. The Fire Spirits began to crackle, sending the humans into a dreamlike state in which they could hear Antelope with their hearts. All was quiet as Antelope rose to speak. "You, Earth People of the Two-legged Clan, are of all five races, and yet among you there is no understanding of survival. Now that our world has become harsh and barren, it is up to each of you to use your gifts of intuition and creativity in order to survive. Still you will be missing the animal instinct, which would teach you how to recognize danger, find food supplies, know the healing plants, or hear nature's teachers. I offer my body as food for you so that through my body you will gain these instincts. I offer my hide and fur to warm your frail frames. My bones will make tools to aid your survival, and my horns can be fashioned into strong implements to allow you to hunt further. The sinew of my tendons will be thread to sew your clothing, and my hooves will make glue if boiled in my belly with water. This is my gift. Others of the Creature People have agreed that you are in need of what we can offer. Honor our lives, our instinct, our bodies, and do not waste these gifts. Act now and you will survive." Antelope had spoken.

After that second gift of life, the first being fire, the children of the Two-legged Clan began to grow and prosper. Through eating animal flesh, they became more aggressive and full of survival instinct. They watched the Creature-beings in nature and mimicked them to their own advantage, learning the secrets of survival in the world covered with blue ice.

Many long, grey winters passed, and my life among the cave dwellers increased in richness. I saw the growth of order among them as they created ritual and ceremony based upon the wisdom they had received from their Totems. The Creature-beings had emerged in the dreams of the human race in order to teach them the mystery of being spirit in an animal body. The women of the Two-leggeds had formed families and then Clans based upon the various Totems who guarded each person of the group. Those who had the same Totem came together in order to understand the teachings of their dreams. The sign language emerged as these Medicine Stories were shared.

The rituals of each Clan were set to music as skins were stretched across hollow logs, creating drums. The leg bones of animals, left from former dinners, became drumsticks used to beat out the rhythms and sounds of various animals. The ceremonies of the Two-leggeds emerged with dancing and storytelling, and they acted out the particular events through mime and sign language.

My life was full in those days as I dreamed with one healing woman after another as I was passed down through the generations into the hands of Alona. She had come to understand many of the healing ways of nature. Each spring brought the new growth of healing herbs nestled among the remains of sparse glacial ice. The seeds, planted many seasons before, had taken hold. My mission among the Two-leggeds was being fulfilled. My seven talents were being understood, developed, and used. The young woman, Alona, was becoming a teacher for others as she began to understand the dreams we shared. The familiarity I felt with her spirit was a reminder of the First World. The love and inner peace she held was a comfort to me as we passed time together. My intuition told me that I had known Alona's spirit before, even though her body was different.

One bitter winter night when food was scarce and the Ice-beings traveled on Raging Wind, who beat the hides that covered the entrance to the cave, a low moaning sound came across the tundra, mingling with the snowy storm. No one paid much attention, as it was nearing the time for the Clan to sleep. My ears perked up, hearing this haunting melody, so lonely and forlorn. Alona stirred and listened. Her curiosity outweighed her fear as she tiptoed to the edge of the cave and peeked out. The storm prevented her from seeing the source of the song, but she waited, huddled in the darkness, until the storm subsided. Grandmother Moon appeared and the song grew in the inky night. There on the tundra was Wolf, white as the driven snow, singing for Alona. Wolf waited, and almost as if she knew the others were sleeping, she began to sing to Alona's heart. Alona knew that her Totem had finally come to her and that the song was a greeting of one spirit to another.

That night Alona dreamed with me, and we sorted out the language that would be taught by Wolf to the Two-leggeds. The Earth Tribe was in much need of a broader way to communicate. The grunts and groans mixed with sign language no longer expressed the fullness of feeling that these Two-leggeds were experiencing. Many had been Sky People who had chosen the limitation of human bodies in order to feel, touch, and share with their hearts. Now Alona would have to complete part of her mission in this Earthwalk by teaching Wolf's language to her fellow humans. Through Alona and others, the brown race became dominant in the Second World, because they used their talents to survive and eventually flourish. Unlike the other races, these brown children of the Earth Tribe carried *the remembering* in their frontal lobes and could access the memories of what had come before.

I was filled with happiness as I reviewed the potential talents that could be developed through speech and through song, which was used by their Creature counterparts. These Two-leggeds were free to create on a level that would allow the beauty of their creations to be shared in a new and different way. Good Medicine was flowing through the Second World of Creation, and I was allowed to be a part of it. Alona felt my joy and cuddled me closer in her hand as she nestled deeper into her sleeping robes. That winter morning we began to share the tundra Wolf's song, and humankind started learning how to speak.

As Alona came to adulthood and had a family of her own, the growth of the Clan expanded. The cave paintings took on new meaning for the people as each artist recalled, through spoken story, the inspiration behind the painted scenes. Life was changing among the brown faction of the Earth Tribe. Each step in the growth process brought new discoveries and creations. It was becoming easier for these children of the Two-leggeds to survive. The tubers, berries, and wild vegetables were gathered and stored in summer for use in winter. The healing herbs and their uses were better understood, which gave further sustenance to the Children of Earth. These Two-leggeds were learning how to interact with the Plant People and use their Green Medicines to make human life longer and richer.

The Clan Mothers and Medicine Women were the guiding forces that served the Two-leggeds. All things were born of woman, and it was understood that the Earth Mother was the source of all the physical needs of her children. As those needs were supplied by the Planetary Mother, the women of the Clans nurtured their families in turn. This was the natural order of life in the Second World. The men took the role of protectors and providers for the families, developing their talents of courage and boldness. The honoring of the Earth Mother spilled from that knowing into the lives of all Two-leggeds as the men honored the women as extensions of the Original Mother of all things. Harmony ensued and the circle of The Faithful was once again complete.

Alona lived many winters while she shared her courage and wisdom with the Clan. It was many generations later that I came into the hands of one of her great-grandchildren's grandchildren, called Ish Na. The life of the Clan had been slowly changing. Carelessness and desire for control of the Clan's destiny had ravaged the good Medicine that had been nurtured for centuries by the early Clan Mothers. I was saddened to see those changes as the people of the Earth Tribe turned their backs on the Sacred Traditions of their Ancestors. The sparkle of harmony had been tarnished as life became less survival oriented for these brown Two-leggeds. The teachings of the Totems were being ignored, and spoken language now contained disdain and critical words of ugliness. The enemy of greed had surfaced again, and with this ghost of the past, self-importance had stolen the hearts of some of the people. The desire for community and equality had been plowed under, and the new seeds of lust for control were growing into thistles of prideful antagonism and resentment.

The winds of change had blown the seeds of love to the Four Directions. Dominance was prevalent in the attitudes of those inside our sheltering cave. The women no longer nurtured the young as they once had. The men were slovenly and demanding of the women, who feared the birth of new children. The memory of all that had gone before was still adding girth to the frontal lobes of the brown humans' brains, which made the delivery of offspring nearly impossible. The pelvic spread of the women did not grow to match *the remembering* being stored in the heads of the coming generations.

Many of the children born to the Earth Tribe came into the world after long hours of labor and at great pain to their mothers. Some of these babies would never utter their first cry. Others would be born only to live a few breaths of life and then travel the Road of Spirit, *Dropping Their Robes* within the space of a few heartbeats. The Medicine Women were confused as to the reason for the difficult births and frequent deaths among their future generations. A Clan was only as strong as its numbers, for much work was needed to survive in the Ice World.

Among the brown race of the Earth Tribe, the ritual of birth and death had created a Tradition that continued from generation to generation. The birth blood on an infant marked the child as a child of The People. The Clan discovered, through hunting and eating various animals, that all Creature-beings also had red blood, which was a sign of being part of the Planetary Family. Those of the Earth Mother's family were to be smeared with red clay and placed again in her womb when the body died. The Red Earth People knew that they would return to walk the Earth Mother, being born of her body, when the spirits called them to be born again.

These people were of the brown race but were called the Red Earth People because they started using red ochre paint made from clay and the fat of Musk Ox to show that they were faithful to the Earth Mother. Since that time, the color red has denoted faith among those who count themselves among The People or The Faithful. These Two-leggeds gladly give the Robe that cloaked their spirit back to the Mother Earth to fertilize her soil so that, through their empty bodies, others may find nourishment. The Green Medicines of the Plant People grow abundantly from the Earth Mother where a Two-legged who shared in this way is buried.

As the seeds of unrest were growing among the Clan, I sought the solace of the woman-child, Ish Na, who had become my new Guardian. The eyes were different, but she carried Alona's Medicine. I was pleased that she heard my voice and that we knew each other's hearts. Life had become a maze of disjointed events with the internal fighting among the Clan members. Ish Na observed the unrest with tormented eyes and a troubled heart. Disrespect for the Elders of the Clan was no longer an internal mutiny. The hostility was overtly apparent as the power struggle tore at the life force of the unspoken laws that had served the Ancestors of the Red Clay People.

Ish Na was an apt pupil when it came to knowing the portents of the nature spirits. She would spend long hours in the spring of the year practicing nonmovement. Like a Stone Person she would sit by my side and gather everything her senses could perceive, storing the information for future use. Silently we would observe the movement of the breezes and the habits of the Creature-beings and the spirits who roamed the meadows looking for Little People to play with. Ish Na followed the ways of her grandmothers, gathering the healing herbs and clays that would mend the cuts and breaks in human flesh or bones. She often talked with the woodland creatures and mended their wounds. Ish Na was a part of the Red Clay People, but she refused to be involved with the jealousy and quarrels that were destroying the legacy of harmonious living that Alona had bequeathed her.

The turbulence of the Second World began to eat away at the basic values of the Earth Tribe. Ish Na was among The Faithful and often listened to the heartbeat of the Earth Mother. The Mother shared the secrets of the coming changes with Ish Na as preparation was made to cleanse the planet of its human parasites once again. Black Arctic Flowers had collected sunlight and warmth, which began melting the glaciers. The ice mountains were slipping, and fast-moving water was filling the streams near the cave of the Red Earth People. One sudden movement of the great ice mountains could destroy the cave that housed the Clan. Anxiety and resentment grew among this band of the Earth Tribe, because water was now dripping through the cracks in the cave walls, causing their dried food to rot.

Still the people carried on their daily activities with no thoughts of the impending doom. The carelessness and lack of respect for the other creatures of nature made the Two-leggeds blind to the obvious habit changes in their animal counterparts. *The remembering* could not serve those whose self-importance had blocked the Knowing Systems of the Ancestors who had created the memories. For those still living in harmony, a new understanding had been added to the memories. This new wisdom was based upon the truth of the memories as well as firsthand experience. The inner knowing came from understanding the voices of nature who echoed the Earth Mother's voice.

Ish Na shared her Medicine Dreams with me. We discussed what the Earth Mother had told her when we were deep in the forest away from the others of her Clan. Vision had reached her senses with powerful clarity. Much as it had occurred with others at the ending of the First

World, The Faithful were being called. Ish Na saw where she was to go when the Earth Mother decided to purge herself. Those who held humility and truth in their hearts began to whisper the messages they had received. Those who were tricked by their own crooked trails laughed at The Faithful. The imprint of pride and self-importance had taught the Would-be Leaders to ignore their feelings and the place where Great Mystery lived inside them. The harmony of heartfelt sharing and equally honoring every Clan member's Medicine was dying.

One evening the Sky Nation took on a mantle of pea green, and colors spurted from the Earth Mother to the Star Nation in arches of iridescent blends of light. The Aurora Borealis had returned. This Whirling Rainbow, the promise of peace for The Faithful, was our sign to retreat into the inner Earth. The Two-leggeds gathered the things they had collected for the journey and waited until the others were asleep. Ish Na held me close to her heart as she lay in the darkness of her sleeping robes. Long after midnight we began our trek to the ancient Rowan, a powerful Standing Person deep in the forest. The Medicine of this Standing Person was protection. We had been told by the Creature-beings to go to Rowan when we were seeking the entrance to the worlds below.

Rowan was patiently guiding others of The Faithful to form a circle around his giant form. Suddenly, as if on cue, the low rumble started and the Earth Mother belched up a roaring growl. The ripples in her body began to move the great ice mountains. The shift had begun. Far in the distance, we could hear the screams of those in the crumbling cave. Even the animals began to cry in anguish at the horrifying sound. The Two-leggeds were paralyzed with fear as Rowan's roots began to move upward like hundreds of Snakes, moving the loose soil and creating ropes for all of us to slide down into the womb of the Mother Earth. The Subterraniums from the inner Earth were guiding those making the descent into the earthy warmness of the tunnels below.

When the last of The Faithful had entered the tunnel system, Rowan's roots covered the entrance and his voice echoed in our hearts. "My spirit will be with you soon," he cried. "Thank-you for the opportunity each of you has given me to serve the Earth Mother and her children. I can see the mountains of blue ice moving toward me. They will make my passing swift and painless. My spirit will fly with Raven into the Void, bringing us together soon." Rowan's voice trailed off into the howling winds of the Second World as the rumble of the great ice mountains passed above.

For days we traveled amid the underworld tunnels, breathing the moist, rich warmth of our Earth Mother's breath. The smell of wet leaves and moistened soil after a summer rain accompanied our journey. Among our group were the Little People, the Two-leggeds, Creature-beings of every kind, Creepy-crawlers of the insect kingdom, other Stone People, and the seedlings of the Plant Nation. Many who could not move quickly were carried by their Brothers and Sisters who could provide easier modes of travel.

When we reached a giant cavern, we stopped to give thanks to Great Mystery and Mother Earth, holding a Council Fire and ceremony of celebration for the safe journey.

Among our thoughts of thanksgiving were the remembrances of the gifts and legacies that had served the Children of Earth in the Second World. We praised the gift of Fire, the forming of Clans and families, the spoken language of the Two-leggeds, and the honoring of the Earth Mother and her prodigy, woman. Thanks were also given for the hard lessons that formed new understandings for the Children of Earth. These lessons were based on how self-importance could destroy the inner balance among all creatures. We were especially grateful for knowing how to work together without ego. This group consciousness brought us unity and wholeness. We spoke with reverence about developing and sharing the talents that could assist everyone. We praised the caution and instinct that had been given to the Earth Tribe by the Creature-beings through the consuming of animal flesh. Everyone gave thanks for the unselfishness of all family members and Ancestors who had given of their life force so that The People could live.

In our celebration ceremony, we gave thanks that the natural systems that ruled the order found in Nature had been coded in light by the Rainbow of Peace. This new natural order had taken root in the Second World. This kind of good Medicine was a reason to dance, to sing, to drum, and to be grateful that those talents had come into being during the Ice World. The Earth Mother was purging herself of those who would not live in harmony. Out of the Mother's love, those spirits would be given another chance to grow in the Third World.

As I looked around the Council Fire, I saw the various forms and faces of The Faithful. Many different Tribes and races were represented. I was feeling the light of inner knowing and trust by being among these who saw their similarities rather than their differences. The circle of trust was complete and represented the wholeness of the Sacred Hoop of the Planetary Family. The lessons of the Second World were understood by every life-form represented at our Council Fire. Once again, our faithfulness had been rewarded, as was promised by Great Mystery in the First World. The Stone People present remembered the events that had promised that The Faithful would always be taken care of. The records of those promises had been put into the petroglyphs that were now shattered. No records in the cave would remain in the above world after this second purging.

My mind skipped back to the various cave paintings that had recorded much of the journey of the Earth Tribe. They were gone now and I was saddened. I began to hear a whisper in my mind and the Earth Mother's voice resounded in my being. "Geeh Yuk, you are called Seven Talents," she whispered. "You have seven sides to your body, and these seven sides represent the Seven Worlds through which the Children of Earth will travel in order to evolve. Feel the fire of my molten core. Allow my love to reenter your body, as it did when your form was forged during the creation of the physical world."

I felt the warming within my rock form as a sudden blast turned my innards molten once more. The intense heat traveled to the surface of one side of my body and etched markings that recorded the events of the First World into the living rock. Then once again, on the second side of my glowing moltenness, the markings of the Second World appeared, recorded for eternity. *The remembering would now be available for all of the Children of Earth.* Deep within the fiery core of my being, the truths regarding my personal journey began to emerge. Slowly, faint traces started to surface of what would come to pass as I fulfilled my Earth mission. In that moment I knew that my mission had been made clear because I had remained among The Faithful. My heart soared with Eagle high above this inner sanctuary upward into the deep midnight blue of space. The freedom of who I was as a Stone Person and the role I played in the this beautiful plan of Earth evolution was etched into my body forever.

There was no more longing within me to be something other than what I was created to be. The Second World gave me the understanding of how to use my seven special abilities. I gazed into the Council Fire and saw my desire framed in the warmth of the Eternal Flame. I looked at the faces of a multitude of different types of Children of Earth sitting in our circle of relations, and I understood the beauty of each original creation. Their gifts and talents represented every talent needed for the Earth Tribe to attain wholeness. Each role was perfect and complete. The Faithful were the Guardians of the Sacred Hoop.

Here, at the end of the Second World, the new beginning
of another world filled with promise was awaiting birth.
Every part of my being knew that the future and the past
were good. Although the Ice World would not really end
until the beginning of the next world, the time in between,
before we emerged again, was a time to give thanks to
Great Mystery for all of the lessons we had learned. It was
a time to see that the difficult and the joyous steps of our
learning process were equal in value.

So ended the Second World of Ice.

CHAPTER FIVE

THE THIRD WORLD OF WATER

The Third World would be known as the Water World among those of us who were the librarians of Earth History. The Stone People, who were the Record Keepers of all events, etched the memories of the first two worlds deep within their beings in order to preserve those worlds' lessons and the truths learned during those times. As each world was purified, the previous civilizations and their cultures were wiped from the face of the Mother Planet.

The Second World had ended, as it had begun, with an Ice Age. Now the one great land mass, which was called Turtle Island, was vast enough to almost encircle our Mother Planet. The mountains and valleys were again growing beautiful foliage, and the warming was supplying plenty of water to assist the young seedlings. I had watched this regreening process since we had come from the realm of the Subterraniums, the Lives-Below-the-Earth People. The human repopulation process, including all five races, was as fruitful as the lush plant growth before me.

The Little People had once again hollowed their homes in the forests and glens and the Creature-beings had

L. Childers

traveled to the Four Directions, covering Turtle Island. Life was taking hold anew in the above world and the Planetary Family was flourishing. Near the heart of Turtle Island was the only Tree Person who had braved and weathered all the changes of the previous worlds. This magnificent Standing Person was called the Tree of Life. Her branches were filled with every seed, fruit, nut, flower, vegetable, and leaf that was present on the Earth Mother. Her roots went deep into the breast of the Mother Planet and took nurturance. The Tree of Life was the Chief of all Standing People and continued to provide the seeds for repopulating the soil after every world's purification.

I came to know this Grandmother who was the Chief Standing Person in the Third World when I was taken to live nearby. The Little People had a special fondness for her, and while giving me a ride they had chosen to make their homes in her root base, which spread for miles beyond her trunk. The Tree of Life represented the Earth Mother's promise to all her plant kingdom children that their gifts would be used by the Creature-beings and the Two-leggeds as a life-support system. The Tree People would provide food, oxygen, lodging, bark paper, healing cures, shade, firewood, glue, dyes, and much more for the evolving Planetary Family. The Tree Tribe gave of themselves totally in order to assist the other Clans of Earth. In return, they were supported by the Mother Earth's energy, which they drew from their roots, Grandfather Sun's loving light, and the Cloud People's rain, which came from the Sky Nation.

Those were happy days as every Clan of Our Relations found a way to be of service to every other band and Clan of the Planetary Family. From time to time I traveled with Condor, held in his talons, to the other areas of Turtle Island in order to see the progress being made. High above the Earth, I could view the exquisite patterns of green and growing Plant Relations blending with the purple snow-capped peaks. I watched the seas crash upon white sand shores and crossed over meadows filled with splashes of rainbow-colored wildflowers. I noted every movement of the herds of Creature-beings wandering through golden plains of windswept grain, feeling that I, too, was a part of the harmonious movement of the cycles and seasons.

Thousands of moons passed and life abundant flourished on the Mother Planet. One day when I was visiting one of the Little People near a waterway that surfaced from the subterranean caverns to divide a mountain valley, I heard some exciting news from one of the Crystal Clan. This Crystal Person, called Gahgwihoh (Shines Clean), was bubbling over with the importance of this new information. I sat patiently and listened as he began to explain the wondrous discovery that the white Two-leggeds had come upon. "The Great Ice Mountains, which carved new waterways above the surface during the last world's final days, have given the Two-leggeds a new idea," he exclaimed. "Their Medicine People call us Does-Not-Melt-Ice or Stone Ice. They say that we may have the Medicine to heal the sickness among their people because we reflect Grandfather Sun's love to them."

I was filled with expectation as Gahgwihoh, Shines Clean, relayed the stories he had heard from others of his Crystal Clan. "It seems that the white race have been shown some of the entrances to the inner Earth and have spoken with the Lives-Below-the-Earth People," Shines Clean began. "They have collected millions of Crystal People, who have offered to line all of the inner-world caverns and deep rivers. Through our bodies, the reflected light of the Earth Mother's inner sun will feed nutrients of various colors into the waters of Earth. Then those waters can be used to prolong the lives of the Two-leggeds."

I was elated that Shines Clean had found out how the Crystal People could be of further service to the Planetary Family. We both held records of the missions of all the five human races, so I commented to Gahgwihoh about the white Two-legged Clan's abilities. "Shines Clean, you know as well as I do that the white Two-leggeds have the gifts of creativity, magnetism, and authority. We must give thanks and honor their paths in using this new Medicine as the part of the first development of those gifts."

On that warm summer's day, little did we know that the *Gagan* race (the white humans) would discover their talents of creativity, bringing strong Medicine and new ideas into our peaceful world. Shines Clean and I watched for centuries as marvels we had never seen before poured forth from the brilliant minds of the white Two-leggeds. They discovered cures for all the human diseases and developed ways to restore their physical bodies' health by using the colors of the light from Grandfather Sun. They Entered the Silence and then journeyed with their spirits, following sunbeams across the Sky Nation to the Spirit World. In the process, they discovered how to harness the Sacred Fire of Grandfather Sun, which would later propel Iron Birds that we called Sun Canoes high above the Earth.

Shines Clean would occasionally bring back the news of the latest stories he had heard from others of his Clan, startling us and making us realize the strength of these new Medicines. The *Gagans* had used their authority and magnetism to harness the energy that followed the Trails of Fire Sticks on the Earth Mother's surface. Every place where a Lightning Being struck the Mother's body, the energy could power their enormous camps.

The *Gagans* developed curious things, which they called machines, to do their work for them. We called these curious forms Makes Work Easy for Two-leggeds. The power behind the Makes Work Easy For Two-leggeds was the knowledge of the Fire Sticks' trails, which began at the top of mountain peaks and traveled to lower outlying areas and through the valleys. These Two-leggeds were adding new experiences to the Earth Records more quickly than we could add the information to the Tablets of Truth.

The Tablets of Truth were not an actual set of scrolls that marked the history of Earth; rather, they were the combined records of the Stone People. Every rock, stone, mineral, crystal, pebble, and boulder on the Earth carried a part of the Tablets of Truth. The wisdom of all this new *Gagan* Medicine was encoded in the bodies of the Mineral Kingdom Tribe. The new machines and Medicines created by the white Children of Earth were clean and preserved the Earth Mother's body. There were no roads that would scar her surface or excavations that would gut her body—at least, not in the beginning.

As moons passed the *Gagans* lived much longer than the other races, causing a deeply disturbing set of events to begin. These white Children of Earth did not share their Medicine with the other races of the Human Tribe and began to feel superior since they did not Drop Their Robes, or die as young as the Two-leggeds of the other races. It was then that the *Gagans* decided to enslave the other four races of humankind. They believed that their new Medicine was strong because it was white and clean; therefore, anything that was not white or pristine was bad and could destroy the power they believed they had found in the color white.

Soon the *Gagans* began to fear losing the authority and power they had in controlling the races of different colors. The fear of dirt, dust, soil, and sand began to erode their sense of belonging to the Earth Mother. This is when the white race earned the name the *Aga Oheda*, or the Afraid of Dirts. Their totally white encampments were made of Marble Stone People and Crystal Mineral People. The trees and flowers were being replaced with plants that would never lose leaves or blossoms, which could dirty the paths to their lodges or communal fires. The only members of other races allowed in the Afraid of Dirts' fine encampments were servants, dancers, or musicians, who would amuse the *Gagan* chiefs at their ceremonies or feasts.

Darkness had crept into our world upon the shadows of fear and had twisted the *Gagan* Medicines, which could have served all races and All Our Relations. In following the Crooked Trail, the Afraid of Dirts began to further demean the other four races by misusing their gifts of authority and magnetism in order to intimidate and control. We of The Faithful watched and felt a great sadness that the new *Gagan* Medicines, which could have brought so much joy to the Children of Earth, were being hoarded for only the few.

At this time all Two-legged Earth Tribes chose to live in areas of Turtle Island that best suited their individual characteristics and in turn aided their survival. The red race built their fires and constructed their lodges in the original center of Turtle Island. The white race resided in permanent lodges to the northeast of Turtle Island's center in an area that had less sun and heat, in order to preserve their fair skins. The brown race camped to the south-southeast and to the west, the yellow race lived near the farthest western shore. The black race dwelt in the farthest southeastern area and the farthest southwestern area, where the climate was hottest and Grandfather Sun's light was strongest.

Turtle Island was immense, nearly encircling the Earth Mother's waist, and the Afraid of Dirts intended to control it all. So great was the fear of infection by anyone or anything that was not white, and therefore, not clean, that the Afraid of Dirts began to cover the Earth's surface with a white substance, which The Faithful of the Planetary Family called Hard-Like-Rock Snow. It was a very curious Medicine, this rock-hard white robe that the Afraid of Dirts wanted to pour on the Earth Mother's body. Fear clutched at the hearts of the red Children of Earth when the Hard-Like-Rock Snow covered the Afraid of Dirts' land. Then the *Gagans* poured the hard white robe over the red race's hunting grounds, covering the burrows of the Little People, driving the Creature-beings from their homes, and smothering the food that the Earth Mother provided for All Our Relations.

Around this time, the Afraid of Dirts' Medicine People found a way to melt our Crystal People relations and to mix their fluid bodies with other ores found on the Earth Mother. The Earth Medicine that would bind the melted Crystal People to metal or ore was found in the southwestern area of Turtle Island. The volcanic caves created by centuries of Mother Earth's eruptions were now forcibly mined by the brown, red, and yellow races, who had been enslaved by the Afraid of Dirts.

The Wanna-be Chiefs in the Afraid of Dirts Tribe forced the other races to take the ores and minerals from their Sacred Spaces or natural homes. The weight of each load of ore removed from the caverns seemed to represent the sorrow we all felt. These caverns had been created by the Earth Mother to house the Lives-Below-the-Earth People. In trust, the tunnels through the Below World had been shown to the Afraid of Dirts. The earlier trust, which had assured the Below-Earth Tribe that the *Gagan* Medicine would be used to benefit all the Tribes of Earth, had been broken, and The Faithful were betrayed and deceived once again.

I watched the beginning of the Third World's decline and I cried inside. Shines Clean was fearful that many of his Clan were being melted and therefore, murdered, in order to aid the Afraid of Dirts Tribe in their desire to harness the Earth Mother's natural forces. We clung to the faith that the Earth Mother would find some way to break this stranglehold of control before the *Gagan* race destroyed the world. There was no balance in the frenzy with which the Afraid of Dirts were now using the resources of Turtle Island without replacing that which they had taken. The rock and soil foundation of Yeo Land, or Woman Land, the western heartland of Yoedzaze, the Earth Mother, was being eroded.

The Tree of Life stood in the central region of Turtle Island near what is now called the Great Lakes, although the lakes were not there at that time. In *Hail-lo-way-an*, the Language of Love, The Faithful were called by the Earth Mother to gather at that site. All five races of Two-leggeds secretly made pilgrimages, coming from the far reaches of Turtle Island to join with others of the Planetary Family. The Creature-beings came and some brought Stone People with them. Other Stone People were carried by the Little People, and many of the Crystal Beings were secretly carried to the site wrapped in leaves. This camouflage was necessary because none of the slave races were allowed to have a Crystal Person in their possession.

When The Faithful were gathered at the Tree of Life, we were forced to confront our greatest fears. The soil around the Tree of Life was covered with the white substance, which looked like the web of a giant Spider, strangling the roots of the tree. We knew that the Great Grandmother Spider who wove the webs of life would not have destroyed her creations. It was evident that the Afraid of Dirts had been there, but there were none currently in sight.

We gathered our wits about us, laid aside our fears and sorrows, and then went to work. Those who could began to chip away at the robe of white. The Mountain Ram Clan battered the Hard-Like-Rock Snow that made the weblike robe. The Little People took out their tools made of Flint and pounded. The Two-leggeds used the Granite Stone People to hammer at the base of the roots, allowing them to breathe, and Buffalo pounded other spots with their hooves. But the white Hard-like-Rock Snow would not give way.

It was time for a Council to be held; we needed to make a decision. We dared not light a fire on the Hard-Like-Rock Snow and give away our location. Instead, we gathered around a coal that was carried in the horn of a great Bison that had been brought by one of the brown Two-leggeds in our company. The Tree of Life carried the seed of all future generations of the Plant Tribe. This web of white would strangle her roots, reflect too much light under her limbs, dry her fruit, and destroy the balance she maintained with the Earth Mother and the Sky Father. We had to do something quickly.

Before we could decide upon a plan of action, the Afraid of Dirts arrived in one of their Sun Canoes. Hundreds of them began to cut the Tree of Life to shreds. The screams coming from the Tree Tribe filled the winds, and a howl of enormous pain began to swirl across Turtle Island. We wailed and grieved with our Green and Growing counterparts, and yet nothing would stop the monstrous actions of destruction that the Afraid of Dirts unleashed that day.

Every one of The Faithful could hear a pronounced whisper that filled the wind. The destroyers seemed to hear nothing. We listened more closely and heard the voice of our Earth Mother as she spoke to us through the Language of Love. "Children of Earth, hear me," she whispered. "I felt this dreadful thing would come to pass. It is up to each of you to gather the seeds, fruits, bits of bark, and roots from the Tree of Life while you can. These Afraid of Dirts can destroy the Tree of Life, but they cannot destroy life itself. The promise of every future Sun and Moon cycle lives in the seeds of today. Take every part of this Standing Person that you can secret away and save the promise of future generations for those who will walk this world after you are gone."

We of The Faithful began to act as one mind, one body, avoiding the *Gagans* as much as possible and hiding what we could that fell from the Tree of Life onto the white robe covering the soil of our Mother's body. The intention of the Afraid of Dirts was to put fire to the remains of the Tree of Life, and their orders included destroying every part of the root system. They wanted no part of the Tree of Life to remain as a symbol of rebellion among the four enslaved races of humans. The fact that they had inflicted so much hurt on the Creature-beings showed that they considered them dumb animals, and so they just herded our Creature Relations away with no real interest. This herding of the Creature-teachers allowed us to gather as much as we could while they spent the next few days cutting through the white robe of Hard-Like-Rock Snow to destroy the outlying roots.

When only the stump remained, connected by a giant taproot, the Afraid of Dirts began to crush the Earth Mother's final connection to the Tree of Life, pulling bits and pieces of the taproot from her breast. It was at this point that we heard Mother Earth's voice once more. "Listen, my children, let this time of sorrow be a reminder of how the Planetary Family lost its original Earth connection. Speak of this day around your Council Fires and relate the dread consequences to those who have ears to hear, eyes to see, and hearts to understand. I will preserve the lives of those who will carry on the Tradition of the Language of Love. The connection to me will now have to be carried in your hearts and in every act of life you perform. The Tree of Life provided sustenance for all the Children of Earth, but now that connection must be sought out by those in the worlds to come, for as a result of this destruction, it is no longer a right but, rather, a privilege."

I questioned the Earth Mother: "Tell me, my true Mother, what would you have me do? I am here to be of service, and yet I feel so helpless."

The Earth Mother answered me with a deep sadness. "Seven Talents, you must take this band of The Faithful away from the Tree of Life, for when the taproot is totally destroyed, water will come from the Below World and cleanse the Afraid of Dirts, who have destroyed Great Mystery's gift to the Planetary Family."

I answered in a whisper, as if I were afraid of being heard. "Great Mother, is it your wish that the Third World end by flood?" She quickly replied, "It is not my wish, Geeh Yuk, but it is what has come to pass. The Two-leggeds who are my children have always been given free will. It was their choice to destroy the roots that kept them connected to the abundance that was their natural legacy. The waters of the tides of change will purify the damage done this day, but the legacy of their wanton destruction will be felt in the coming worlds and will be inherited by all future generations. You must hurry and gather The Faithful, for the time is near. This act of murdering the Standing Person who has given them the seeds of life will begin the floods that will always come when the roots of the Plant Relations can no longer be the bridges between the Earth and Sky Nations. Hurry, my friend, and do not fail. Take the others to safety."

I spoke to the Little Person at my side and we called to the others to follow. As we traveled far and fast, we were troubled by the impending events the Earth Mother had spoken of. A few hours later, we stopped to rest, and we were chilled by the cries of anguish that were suddenly carried on the Winds of Change. Hawk screamed for all of us to look high in the Sky Nation, and there we saw a geyser of water many miles back reaching toward Grandfather Sun. We decided to head for higher ground in order to find the safety we had been promised by the Great Mother.

A rumble came from underground; the Earth Mother shook violently for a moment and then was still. Water was still spurting from the geyser, climbing skyward. The small band of The Faithful joined together to find the fastest way to higher ground. Many of us who were tiny were carried by Condor to the safety of those mountain heights.

When all were safely gathered, we heard the Earth Mother's voice once again: "There will be new sources of water in the area where the Tree of Life stood, my children. There will be Great Lakes, which will serve as a reminder of the deeds of this day. The flooding will continue across Turtle Island every time the sacred bond of life and equality is broken by the Two-leggeds. It is time to call all the Medicine Stones together, because they are the Record Keepers of the true history of our journeys together. There, high on Sacred Mountain, the truth of the coming worlds will be revealed to The Faithful through the records of the Stone People. The Council Fire you will hold will decide and proclaim the ways in which The Faithful may be reminded of the Pathways of Truth and Earth connection in the future."

These words were held in the hearts of The People as we completed our tasks to preserve the alternative connections to the Great Mother for the future generations. Over the next few thousand rotations around Grandfather Sun, we watched as the Afraid of Dirts' mining and greed eroded the foundations of Yeo Land and made that part of Turtle Island sink as the waters took it over. The great blue seas then claimed the brown race's land to the west. Many hundreds of moons later we saw the Afraid of Dirts destroy their own part of Turtle Island, called *Gagan* Land or White Land. At the end of the Third World of Water, The Faithful went once again to live below the surface of Turtle Island as the purification was completed.

The destruction of the Planetary Family's original connection to the Earth Mother had set the stage for the Fourth World, which would be called the World of Separation. The unity of Turtle Island was no more. At the end of the Third World, Turtle Island was split, and new continents were formed from the one great land mass that had been our common home. The waters of purification separated each regional land mass and split the areas where the five races of the Human Tribes had once lived as one. This legacy of separation would change the way in which the Planetary Family perceived the past and the future. The future lessons would be difficult, but life would continue on the Earth Mother. The compassion of the Great Mother came into being as she bequeathed to the Children of Earth another chance to grow through the ways of separation.

So ended the Third World of Water.

CHAPTER SIX

THE FOURTH WORLD OF SEPARATION

The Fourth World gurgled and spurted its way into existence as the waters receded and the steam from volcanic tide pools returned to live in the Sky Nation once more in the form of Cloud People. The previous Water World had left its mark on the Earth Mother's body, and the cleansing was now complete. The memories of the Wanna-be Chiefs or Would-be Rulers of the masses had melted into the oblivion of the deep. Below the oceans, new life stirred and filled the vacant remnants of the old civilizations that had gone to live beneath the waves.

It was now time for the remaining Children of Earth to emerge and take their rightful places among their Relations above the soil, to begin again and to become the new foundation of The Faithful. Much had been learned about the natural order of our planet during the demise of the Water World. *Promise* was the watchword among those who chose to repopulate the above world, because the Earth Mother promised the Tribes of Earth that they would never again live under the domination of one race. The new lessons of humankind were to be riddled with separations even among the Clans of individual races, and therefore, no single race would ever be able to agree among themselves enough to exercise control over others.

As it was in the beginning of every new world, the promise of the Fourth World was the harmony of equality, but the challenge was of separation. Since it was time to begin at a new place on the Medicine Wheel, The Faithful carried this promise of equality as well as the challenge.

I watched in wonder as those who emerged with me from the lodges of the Subterraniums made plans for the future. Life was once again fertile and abundant. The Little People set out to reseed the soil in every hill and valley. The Winged-ones ate great quantities of seeds and flew across the land depositing fertilized pods to the far reaches of the horizon.

As life took hold and new generations of humans and Creature-beings were born, the new order was set in motion. This new order of learning how to see differences and uniqueness would later deteriorate and become known as the Fourth World of Separation. The members of Planetary Family were beginning to understand their separate missions, their distinct species, and how each spoke on the Great Medicine Wheel could interact with all others and still retain its identity. Special talents or Medicine in one given area could create a support system, and yet the memory of the destruction of the formerly unified Turtle Island loomed in the shadows as a grim reminder of how Medicine for a select few could create disharmony and pain.

The remnants of Turtle Island floated above the blue seas of seeming contentment as the new generations refused to look deeply into the watery past, which had left them with the legacy of separateness. Each emerging continent was filled with different traditions, cultures, and ways of speech. *Hail-lo-way-an,* which had been the unified Language of Love, was all but forgotten. New languages were taking the place of the primordial tongue, just as all races of the Two-leggeds were segregating themselves in order to learn more about their individual makeups. I could see the value of specialization, but I dreaded the future wars that could be fought in the name of only one philosophy being holy or correct.

As the Two-legged children of our Planetary Family began to explore their individual beliefs, the understanding of the concept of Great Mystery dwindled. All concepts that contained an expansive overview were disappearing, and pieces of the whole idea were taking their places. As time moved steadily forward, the intellects of the Human Tribe began to conquer and override the unified concepts that had represented oneness in the former worlds. Symbols, originally endowed by Great Mystery, that had contained the concepts of wholeness were lost when new ideas containing bits and pieces of truth were adopted by all five human races.

The individual races of humankind began to form religions based on limiting beliefs, each religion being controlled by Wanna-be Chiefs who basked in the seeming knowledge of the new belief systems. Throughout the passing centuries, spirituality as a way of life and a way of being was replaced with sets of laws based on fear of angry gods. The Stone Tribe dutifully recorded each passing change as the various civilizations peaked out and then collapsed when their lack of equality finally created division from within. As the Great Mystery was forgotten by most of the Two-leggeds, many gods were created to take the place of the Original Source. This final and ultimate form of separation from the Original Source was frightening to The Faithful among the Planetary Family.

The Medicine Stones remembered the Council Fire of the Third World and the prophecies that had spoken of these times in the Fourth World when the idea of many gods would lead the Human Tribe through many difficult lessons. We imparted our memories of these prophecies to those who sought our counsel in order to relieve the troubled hearts of the loyal and trusting Two-leggeds. The Fourth World was the longest of all the seven worlds and, being the middle world, would be the turning point in the spiritual evolution of humankind.

In order to evolve with the Two-leggeds and to support their painful growth cycles, the rest of the Planetary Family would need to endure the harsh lessons taught by ideas of separation. It was not easy in those times of uncertainty and change. At some points in this cycle of change, the Creatures of the Earth were worshiped as sacred gods by certain religions, and then later those same Creature-beings were defiled. Some factions chose female humans as gods and others chose male humans, while others still named parts of nature as images of their gods. Each new ruling culture learned from the failures or successes of former civilizations, but in the end, all Would-be Rulers were eventually ruled by their own greed and their lust for power and control.

Wars that killed much of the human population battered and conquered each culture in the name of its own gods or in the name of being the only chosen race. As blood was spilled in sacrifice, poverty and pestilence were the tragic companions that followed the crooked trails of power and glory. The destruction of brotherhood and sisterhood among the Two-leggeds broke the bonds of the heart. Material possessions were used as symbols of success that could impress some of like mind or intimidate others who saw their own poverty as a symbol of their unworthiness. The jaded senses of the Two-leggeds lost connection with the beauty found in the natural world and led these confused Children of Earth down trails filled with demons of their own making.

Our hearts were heavy as we watched the burning of their great cities, the destruction of their libraries of knowledge, and the human wreckage left behind in their holy wars. The Faithful still kept the Eternal Flame burning brightly in the below world of inner Earth. The Sacred Hoop of all nations and all life remained intact as it had in all the former worlds, safe in the keeping of the Guardians and Ancestor Spirits. As long as The Faithful had life and breath, the Sacred Hoop of life would remain whole. The members of the Planetary Family who had lost their true connection to the Great Mystery projected their own sense of separation into the world by whispering that the Sacred Hoop was broken and needed to be mended. Little did these broken spirits know that The Faithful still stood watch and maintained the connection to all life in order to give

The People, who might have briefly lost their faith, a chance to return to Great Mystery's Medicine Wheel.

As time moved like the rolling rivers past the curves and bends of experience, the attitudes of humankind were shaped by their fear and loss of Earth connection. Many of the emerging philosophies and religions were kept busy degrading the natural actions and functions of being human. Humans felt it was unsafe to use creativity and self-expression, because they feared being too different. The limiting thoughts of separation cast doubt on those who heard a different drummer, making fear and hesitation's victory over the Two-leggeds complete.

The sacredness of sexual union was defiled with polarizing ideas that created rules concerning how each human, depending upon gender, was allowed to express sensuality or desire. The stricter the rules of various Clans and Tribes became, the more often sexual crimes were committed. Dogma and morals replaced the inner knowing that had been present among humankind in the former worlds. The fear of being labeled as bad or wanton kept the Two-legged females from expressing the natural body rhythms that maintained their connection to the Earth Mother.

The male humans began to seek companionship with women who were pure, in the judgmental estimation of their newfound religions. Then these men found that many times, the woman was not able to respond in a natural way due to her fear of her own hidden or suppressed sensuality. The Fire Medicine of passion and spontaneity was being separated by the ice that flowed through the veins of all humans who feared their feelings or their natural states of physicality. On the other side of that lesson were those who had lost all connection to the spiritual and were controlled by their own lusts and jaded needs for sexual domination.

The Medicine Stones, Creature-beings, and Plant People watched the changes with trepidation, feeling with utter dread the confusion of those who wanted to be liked or admired. With their human need to be needed, the seeds of self-doubt had taken hold and the roots of separation had strangled the Knowing Systems that had once given them the instinct of their rightful places in the physical world. The pain of the Two-leggeds' lessons was very difficult to observe as they wandered further and further from the Path of Beauty. The changes in understanding that would heal the Human Tribe seemed to be very slow in coming, but the promise of that illumination lived in the prophecy of the last days of the Fourth World. I was troubled about the events I had recorded as each new horror of further separation took hold, but my vision was clear and I nurtured the promise of change that could result in enlightenment for the Planetary Family.

Near the end of the Fourth World I came into the hands of a spirit I had known before. This part of my personal journey had been revealed to me long ago. I had known that Alona, my friend from the Second World, would cross my path once more. I had seen the vision of our reunion many times throughout the ages and was pleasantly surprized as the events from my visions began to take root in my physical experience.

The long green blades of grass gently waved in the wind as I looked into the face of Grandfather Sun. I was admiring a giant Dandelion who had just released a few of her feathery seeds into the gentle breeze, when suddenly an enormous human foot kicked me. I heard the yelp of this Two-legged as he reached down muttering and then yelled at me for being in his path. In a flash, my vision that contained the prophecy of this moment urged me to scream back as the human began to throw my stone body far from his sight. "Don't throw me away!" I yelled, "I am your friend." The human stood totally still as a curious look crossed his face. He shook his head as if he did not believe his ears. Again I spoke, to make sure that he had no time to doubt what he had heard. "Running Deer, I am your friend! I have much to share with you. Please, you must take me home and dream with me."

Running Deer was startled and then amused, and then a look of contentment transformed his features into genuine happiness. "All right, Stone Person," he said, "I'll take you home with me and we will see just what kind of Medicine you have to share." Dragging his leg and moving with great care, he proceeded with me, hobbling toward his cabin on the Cattaraugus Reservation, and I sighed a sigh of relief. As we made our way slowly to his home, I began to know how I could help this fine runner who had been injured. I could feel Running Deer's sense of emotional pain, for no longer could he run with the wind or be a source of pride and accomplishment for his people.

That night, as I lay nestled in his weather-worn, broad palm underneath his pillow, I entered his dreams and spoke to him through *Hail-lo-way-an*, the Language of Love. I told him that I carried a message from the Standing People. They, who could not walk or run because their roots were deep in the Earth Mother's breast, understood his frustration and were sending him a new vision of a way in which they could support and help one another. The Standing People wanted to be of service to other Two-leggeds who could not walk with ease. They sent Running Deer a vision of how their branches could be fashioned into canes in order for their wooden limbs to be of service. In this manner, a part of the Tree Person could walk with the Two-leggeds, giving the human a third appendage for support. Then both human and Standing Person would be enabled to travel, seeing new and different parts of the Earth.

The night was filled with heartfelt messages of oneness, support, and creative vision, and with a prophecy of another kind, as I released my records of these wonders into Running Deer's mind. The silvery light of *Nisa*, Grandmother Moon, pierced the threadbare curtains of the cabin and brought the magic of this shared wisdom to light. The Great Star Nation shifted and moved as Mother Earth turned, and then the glory of early dawn graced the cabin walls with Grandfather Sun's golden light.

Running Deer lost no time in seeking the company of Moses Shongo, the Medicine Man and wise friend who lived next door. Running Deer told Moses about his dream and how he intended to use the branches of the Standing

People to assist his Tree Relations in walking with their Two-legged counterparts, and then he approached the curious part of his wonder-filled night. As Running Deer held my rock form in his hand and offered it to Moses, he spoke of a prophecy that would affect the lives of the Shongo family.

"This Stone Person brought me a dream of a girl-child who would be born into your family, to your daughter Blue Flower. The child will be a great teacher among our people and this rock will be her Medicine Stone. The wisdom she needs to walk this Earth will be given to her through the records this Stone Person carries," he said.

Moses Shongo took me in his generous palms and held me to his heart. We connected our minds and felt the joy of being in one another's company. He looked deeply into the markings etched on my body, felt the Medicine I carry, and spoke words of gratitude to the Great Mystery. Running Deer was pleased that he could be the bearer of good tidings and that his friend would be given the gift of a grandchild. They waxed eloquent as they sat and communicated about the marvelous possibilities the future held in the form of the child to come. Moses's obsidian eyes sparkled as he rocked to and fro in his favorite chair, adjusting his fine felt hat with the beaded band more comfortably over his raven hair while he murmured, "It is good, it is good."

I held a place of honor in the Shongo home as time passed and Blue Flower grew with child. On her birthday the labor began, and she delivered a girl-child who was called Yehwehnode (She Whose Voice Rides the Four Winds). Her name in English was Twylah, and her destiny was held in the loving care of her grandparents.

I watched from my place high on the mantle above the fireplace as the infant grew. I honored her spirit, which was the same spirit as that of my dear Alona of long ago. When Yehwehnode was three years old, she choked and nearly died. Grandpa Shongo saved her life by giving her mouth-to-mouth resuscitation. Moses left the house when his granddaughter was out of danger and walked out to the small cabin that was his Medicine Lodge.

When Grandpa Shongo returned, he called the family together and spoke of the further prophecy that would shape Yehwehnode's life. "Her breath is now my breath and she will carry on with the teachings where I leave off," he proudly pronounced. "People will come to Yehwehnode from all over the world, and they will beat tracks to her door and climb through windows and over walls to receive the words of wisdom she will gladly share. Her life will contain many hard lessons as well," he sadly continued, "for she will be deaf and then she will hear, she will be blinded and then she will see, she will be crippled and then she will walk again." The hush that followed Moses's last words was filled with the individual feelings of each family member as they looked to their hearts for understanding.

Little Yehwehnode would have much to conquer in her life, and her loving grandparents had to teach her much in order to assist her in meeting those challenges. When she was three, they began to teach her how to listen and retain

what she heard, as was done with all Native children raised in Traditional family environments. In that third year of her life, the lessons progressed until we came to the time for our formal introduction. On that special day, the family living room was the stage for a wondrous event.

Many baskets of various sizes were placed on the floor and filled with natural objects representing the different types of Medicine given by each Relation within the Planetary Family. I sat with other Stone People in a basket, awaiting the appointed time. Other baskets were filled with flowers and herbs from the Plant Tribe, bones, fur, and claws from the Creature Tribe, and various Medicine objects that would give the grandparents an idea of what special Medicine Yehwehnode carried. The little girl was placed on the floor and allowed to investigate the various baskets until she found the one that drew her attention and held her interest. Yehwehnode looked over the lot and then came to my basket. My heart soared with Eagle when her tiny hand clutched me to her breast and she refused to let go. Like billions of shooting stars, her sparkling eyes showered me with love, and I knew that we were destined to be lifelong friends once more.

As the winters passed I watched Yehwehnode's grandparents teach her the truths of her people. On the back porch, she would be placed on top of apple crates so that her eyes would meet the eyes of her grandparents. This was to teach the little girl that she was equal. She learned that she was worthy of having her own ideas and her own solutions to the questions presented in the lessons she was being taught. The grandparents were allowed to teach her for only four short years before the government would insist that she go to boarding school and learn the ways of a culture had lost its connection to the Earth Mother.

Yehwehnode learned how to listen and pay attention. She was taught many lessons about the Cycles of Truth. She learned how to use her talents to keep strong in the alien world she would eventually have to enter. One day she took a big spoon from the family kitchen and went to the front of the house to dig herself a hole under the roots of a large tree. On her own, little Yehwehnode had decided to find her Earth connection and to enter the womb of the Planetary Mother. This hideaway beneath the knotted roots would become her favorite place of retreat in those early years.

Grandpa Shongo came to see what she was doing and praised her work, for she was using her creativity. He knew that it was time to present her with a lesson. "Good work, little girl," he smiled. "Are you making yourself a place to smell the Earth Mother's rich breath?" Yehwehnode answered, "Yes, Grandpa, this is the place where I can go inside and feel the Earth all around me. I come here to lie in the leaves and smell the perfume of the Earth Mother." Moses Shongo was pleased, and he answered, "Yehwehnode, where do you think all that soil you have removed is going to live now?"

Yehwehnode had not thought about the soil being moved from its Sacred Space or of having to find it a new home; so she sat and thought. She had to come up with her own solution, as always, in order to develop her reasoning ability. She decided to get a bucket and take the soil she dug up and to go to every other Standing Person and ask if the Tree People would give the displaced earth a new home. This pleased her grandfather, and he smiled on her, showing that she had done well. The approval of her grandparents was not always easy in coming, and she was taught the other side of the lesson in many other instances. The lessons with her grandparents were used to give the little girl strength and the ability to make her own choices without being swayed by the thoughts of others.

The officials of the school could be put off only so long. They came to collect Yehwehnode more than once and constantly asked when she was born in order to find out when they could legally take her from her family. Grandpa Shongo would look at Grandma Shongo and muse, "Now let me see, wasn't she born the year the pump froze, Grandma?" This frustrated the school officials more than once, but finally they refused to take any more vague answers to their questions, and Yehwehnode was taken from her loving home.

During her absence, Yehwehnode's Grandfather *Dropped His Robe* and walked into the Spirit World. Her tiny heart was shattered by separation once again. Far away from those who loved her, in an alien world that refused to honor the natural ways of Earth connection, she cried tears that could have filled a salty sea. Grandpa Shongo's breath was now truly her breath, and at nine years old, she began to shoulder the responsibility of walking in his footsteps and carrying the Teachings where he left off.

I sat in my place of honor high on the mantle and watched the lives of this family of The Faithful as many winters passed. I had been Yehwehnode's playmate and friend during her youth, and now she was married with near-grown children of her own. She had passed the experience of deafness and through her faith had regained her hearing, having learned how to hear through rhythm and vibration in a world without sound. Later she had lost her sight, and the blindness had taught her to see with different eyes, which saw through the illusion of the physical world. It was then that she had developed the gift of second sight, and when she had learned how to use that talent, the physical blindness had left.

Yehwehnode had been also been crippled from the waist down during the difficult birth of one of her four children, and my heart was nearly broken as I watched her fight to walk again. I marveled at the strength of the human spirit and learned much about my own willingness to continue my own spiritual lessons. The lessons she learned taught her perseverance, positive thinking, and further faith in the Great Mystery's plan. Even though her Grandfather's prophecy had spoken of her making it through these difficult lessons, she knew that there was no guarantee if she was not willing to test her own talents and abilities.

The day came when she took me from my resting place and we began to record the Language of the Stones. Each marking on my body was carefully drawn, and I shared the meaning of each symbol in order to make the knowledge available to the Children of Earth. It took us many years using hundreds of others from our Stone Tribe to collect and record the Medicine symbols that composed the Language of the Rock Clans. Yehwehnode would sit with each symbol and Enter the Silence with me as we discerned each concept and meaning.

Together we realized that the Fourth World of Separation was nearing an end and that the ancient tools that would heal the distance between the Stone Tribe and the Human Tribe must be made available. The Language of Love, *Hail-lo-way-an*, was made of concepts that had been destroyed by the specialization and categorization that had riddled the Fourth World with polarities and opposition. I was elated that all of my seven gifts were being used to assist this process of returning the language of my Stone Tribe to the Children of Earth to further our common growth and unity.

I shared my wisdom and my records with Yehwehnode during our years together as we explored how we could better assist the Human Tribe in their growth and spiritual evolution. Many Two-leggeds had lost their connections to the natural world; these connections, if regained, could restore the understanding of the unified spirit of life's cycles and their personal, human roles on the Medicine Wheel.

I shared the records of the Fourth World's prophecy with Yehwehnode one day when the two of us were alone. "The time has come for me to share with you the prophecies that were shared at a Council Fire of the Stone Tribe at the end of the Third World," I began. "We are nearing the time when either the Fourth World of Separation will mark a change in the human consciousness on our Mother Planet or once again, the Earth Mother will rid herself of those who forget to give praise and thanksgiving for the gifts they receive. The First World was destroyed by fire, the Second World by ice, and the Third World by water, and the Fourth World will be destroyed by polar shift or rotation if necessary."

Yehwehnode nodded as I continued, "The purpose of the potential polar shift would be to mark the end of this middle world with a new turning point that would represent the return cycle leading to completion. This shift does not have to occur in our physical world if enough of the members of The Faithful come into alignment and harmony. Each part of the Planetary Family is awaiting the decision of those in the Human Tribe. The center of every life-form has Great Mystery in it. These Vibral Cores send out rhythms of harmony or discord. If the Human Tribe can muster enough of its population who will come into alignment with the rest of the Planetary Family, the polar shift will be avoided."

Yehwehnode sat silently and then shared her thoughts with me. "Geeh Yuk, I can see the awakening of the human Children of Earth coming. I feel that they will meet the prophecy's challenge and move past the ideas of limitation, doom, and gloom that have been created in the Fourth World."

Yehwehnode's words came true as we watched the winters pass. Each cycle of the seasons brought further proof that different groups among the Two-leggeds were joining The Faithful who sought world peace through celebration and gratitude. After a time we were told by the Earth Mother that the Planetary Family had avoided the polar shift through coming together as one.

The Medicine Stones know that we are to experience the transition of the Fourth World and that we will not experience the dawning of the Fifth World until around the year 2013 as the Two-leggeds reckon time on their calenders. It is during this time that the Earth Mother will begin to wobble and send cataclysms to any areas where the Two-leggeds have chosen war, inequality, wrong action, or disharmony. The human decisions for peace have aligned with the Earth Mother's heart.

The Stone Tribe sees the cycles move once again as every member of the Planetary Family feels the wobble that can manifest as life's challenges or difficult lessons. Those who were the Ancestor Guardians of the Earth Tribe knew that the conflict was never outside of themselves but was always within. The fear of having to look at their own inner conflict led many to try to save others who they thought were in peril of losing the only way to enlightenment. This is one of the lies of the World of Separation, for nothing is ever apart; all paths are sacred. All life dwells within the Eternal Flame of Great Mystery and forever lives within all life-forms.

The present Guardians of the Sacred Hoop realize that the exciting adventure now unfolding is one of discovery. The task at hand takes further development of every Relation's talents. Each Relation is being given the opportunity to seek out any conflict within the Self and heal it in order to stabilize the wobble that keeps us from personal balance and world harmony. We must still pass through our personal fires or tests of strength and the planetary changes that will prepare us for the Fifth World of Illumination. Every member of the Planetary Family will be writing that history, and we of the Stone Tribe will be recording it for the future. Now it is time to look at the prophecy of that future and align ourselves with our common vision of peace in order to fulfill the promises of worlds to come.

And so we are now living the final days of the Fourth World.

CHAPTER SEVEN

THE FIFTH WORLD
OF ILLUMINATION

To properly tell the stories of the Fifth, Sixth, and Seventh Worlds, we must return to the moment in Earth's history when the Planetary Family lost its Earth connection because of the destruction of the Tree of Life. Many of the Stone Tribe, who had heard the Earth Mother's words, decided that it was time to hold a special Council Fire before the one land mass on our planet was divided. Underground, the crystal-lined waterways that connected the areas inhabited by the Lives-Below-the-Earth People were crumbling, and we were sure that it would not be long before we would be separated.

The Little People gathered together those of us who would be spokepersons for the Stone Tribe, bringing news of the Council to the others who waited patiently for our return. The Little People traveled far, carrying thousands of their Rock Brothers and Sisters in every conceivable type of carrier. Through the far reaches of Turtle Island, we were brought to a central meeting place high on Sacred Mountain.

Each Stone Person braved the secrets of the Void and traveled in spirit to the dwelling place of one of the Three Fates, Future. As always, the Rock Clan worked as a whole. We were each given a part of a puzzling prophecy by Great Mystery before coming into form. In journeying to the Void, we were given the tasks of gaining clarity on our pieces of future prophecy. Every Stone Person present was chosen to meet the challenges of entering the Void of the unknown. We all Entered the Silence and then journeyed to the Spirit World until each one of us had reclaimed the part of the future prophecy that was reflected in the markings etched on our individual rock bodies.

Many of the visions we received were curious to us, since we had never before seen or experienced the types of activities or the material objects that would be part of the future. I sat silently as our Stone Tribe Chief, Shee-na, mused over the pieces of prophecy we had retrieved from the Spirit World. The Council Fire crackled and roared, breaking the solitude and darkness of the night sky. The curiosity and anticipation among the Council Members was heightened as each minute passed.

Finally, Chief Shee-na spoke. "These visions must be put in order so that we may know how and when they will come into being," she said. Later, when Grandfather Sun had crossed the Sky Nation four times, we had finally completed the task of sorting and piecing the facts together. Grandfather Sun was setting, lighting our circle with his Sacred Fire. We watched Grandfather's light fade to apricot as the Cloud People who guarded the skies above Sacred Mountain turned from white to golden and then to musty rose. The Evening Star, Wata-jis, appeared to guide us, sparkling in the cornflower-blue sky, which rapidly changed to softest indigo. The Little People scurried to and fro adding fuel to the Council Fire. The dried and broken limbs of the Standing People, which were collected from the forest floor below, caught fire immediately, casting shadows on the curious faces of those who gathered in our undivided circle.

The voice of the Stone Chief rang high across the foothills and echoed down the plains as she recited the prophecies of the end of the Third World and Fourth World, yet to come. Long after midnight when the Star Nation was at its brightest and the half-face of Grandmother Moon had set, Shee-na came to the prophecy of the Fifth World. Relief was present in her voice as she told us that when the World of Illumination arrived, the trials, lessons, and sometimes disastrous actions of humankind would no longer have to be inflicted upon the other Children of Earth. I sat in my position in the Medicine Wheel and sent gratitude to the Great Mystery for the words I heard spoken. The promise of the Fifth World was enlightened living that would bring the unity that had been sought for many millions of moons. I brought my attention back to the Clan Chief as she began the prophecy.

"The Fifth World will begin around the year 2013 and will be called the World of Illumination," she stated simply, making sure that she had the attention of all Council Members. "As humankind comes into balance, the Knowing Systems that will guide their growth will be in place. Every human thought has always carried authority and has reflected the Planetary Family's inner desire to create new realms of experience. As Record Keepers, we Stone People have made note of the choices of the Two-leggeds that have contributed to their growth as well as to their more difficult lessons. We are being asked to have compassion and understanding for the Human Tribe's lot in life. During the Fifth World many of the Children of Earth will ultimately find the long-sought enlightenment, the search which has brought them to walk the Earth time and again." Chief Shee-na looked meaningfully into the faces of each Council Member, making sure that we all understood that no disgust for the Two-leggeds would be allowed. Then she continued.

"The World of Illumination will bring a millennia of perfect peace for those who have come to know and use their talents to aid all of the Planetary Family. These Two-leggeds will be called the Rainbow Tribe, for they are the product of hundreds of thousands of years of melding among the five original races. These Children of Earth have been called together to open their hearts and to move beyond the barriers of disconnection. The Medicine they carry is the Whirling Rainbow of Peace, which will mark the union of the five races as one.

"The Stone Tribe has watched these Rainbow Warriors of the Two-legged Clan throughout time, and we have found them among the faithful. These Warriors of the Rainbow are both male and female and are from every walk of life. They will live in every location on the Earth Mother and they have one common characteristic: their ability to live in peace with who they are, with each other, and with All Our

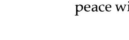

Relations. They are not Wanna-be Chiefs. These Children of Earth do not want to leave the Earth for another home. They are here to dance the Sacred Dance of Life on their Mother Planet, so they do not run from themselves or their chosen paths." Shee-na smiled to herself, recalling the faithful human friends she had known, and then continued.

"At the end of the Fourth World, many Sky People of the Two-legged Earth Tribe will leave the Mother Planet when the Great Medicine Wheel turns, bringing the rebirth of Earth. The Sky People who will choose to leave our Planetary Family have not yet mastered the limitation of being physical. They miss the abilities of limitless creation they once carried in their previous nonphysical forms and want to return to the Great Star Nation. Others of the Earth Tribe will perish in the planetary changes that will occur at the end of the Fourth World of Separation, as they have done in every preceding purification between worlds. Those who will perish have not honored the planet, her people, her creatures, or themselves.

"The Earth Mother's new form will be the realized visions of the artists and Dreamers who have nurtured the dream into being. The Faithful who will remain to create the Fifth World, here on Earth, will be rewarded with new abilities based in strong Medicine that will be used for the good of all." Our wise leader had become excited as she read the previous statements. Then as she read further, her voice took on the timbre of a mountain brook, bubbling with happy harmonies.

"The community of the Fifth World will include All Our Relations in the Planetary Family. Five traveling stars made of Sacred Fire will pass near the sky path of our Earth Mother at the end of the Fourth World. These Stars-with-Tails or Comet People will provide new thought food for the Planetary Family as well as for our Earth Mother. The Two-leggeds will see truth through new eyes. This new way of understanding will stop the confusion that has limited the Two-leggeds' growth. One among the five Stars-with-Tails will fertilize the egg of our Earth Mother.

"The Four Clan Chiefs of Air, Earth, Water, and Fire will be purified through a new type of Medicine brought by the Stars-with-Tails. The Plant Tribe will multiply in this fertilization process and will reclaim the barren deserts. This new Sacred Fire of the Comet People will bring under-standing to all relations within the Sacred Hoop, and the Sky Nation will be married to the Earth World. When these two worlds become one, the Medicine of the Whirling Rain-bow of Peace will belong to the Children of Earth."

Before continuing, Shee-na asked for a moment of silent gratitude so that all among us would know that the trying times of the Third World would have a reward one day in the future. She then said, "All bad Medicine that has been placed in the Earth or dumped in her oceans will be removed, allowing the Great Mother to breathe again.

"Many of the plant seeds from former worlds will be brought to the above world by the race of Subterraniums and replanted on the surface. These ancient varieties of plant life have many healing qualities and will support a more natural way of life for the Rainbow Tribe. Some of these Plant People have not been seen in the above world since the First World of Love. The Subterraniums will begin to teach the Rainbow Warriors how to travel through their underground tunnels in order to repopulate different areas of the Earth. This union of the Subterranium race below and the races living above the Earth's surface will signify and reflect the inner and outer harmonies found in each life-form as well as in the Earth Mother.

"All five races of the Two-leggeds will act as one mind, one heart, and one family, fulfilling the Rainbow Prophecy of the Fourth World. The idea of learning through opposites will be replaced with a Knowing System that sees all lessons as equal. Time will no longer limit our world, and the Children of Earth will be able to use their talents to travel to the Spirit World and beyond. Since the Sky Nation and the Earth World will be unified, the Dreamtime and Spirit World will be easily accessed by All Relations. When this new understanding takes hold, there will be no more illness or death among the Two-leggeds or Creature-beings. Any type of healing that is necessary will be accomplished through the use of good Medicine and feelings. The Dream-time will be as real to these healed humans as their physical reality is in the present world.

"In the Fifth World, individuals will learn in a new way by touching and feeling the lessons needed for their growth. This will be done through developing sensitivity using the body, mind, heart, and spirit as one unit rather than parts of a whole. The art of actually living the dream will be learned through touching and feeling the ideas contained in those visions. Humans will begin to learn through touch and feeling when the Nine-pointed Star of Undivided Truth comes into the dreams of humankind.

The barriers born from wounded hearts, which formerly inhibited the Human Tribe from fully using their talents, will become no more than an ancient memory."

Upon hearing Shee-na's words, our hearts were gladdened for those Two-legged Brothers and Sisters who would one day know the joys of becoming the wisdom rather than seeking it outside themselves. We continued to listen intently as Chief Shee-na went on.

"The Fifth World of Illumination will enlighten the Children of Earth as they see through the illusions that were formerly created by pain. The ideas of truth that have been lost or hidden throughout the cycles of seasons will be sought and found through the desire to know the truth.

"The connections between all life-forms will be seen, and the Divine Plan set up by Great Mystery will unfold. Relationships among all life will become an inner knowing. There will be no more need for the word *universe*. The circle paths of all stars and planets will align to become a Unified World of harmony. The new Uni-world will then create further patterns in the blanket of unified truth throughout all Creation.

"The Stone Tribe will be recognized as a living life-form library, and all markings within the Medicine Stones will be seen as symbol-legends, easily readable through touch and feel. Our healing abilities will be used in order to establish further Earth connection as humans seek the Medicine of their true Mother. The markings on our rock bodies will be seen as intuitive ways to seek truth. The Stone Tribe will teach how to read the true history of the Mother Planet." With this piece of joyous news, a shout rang out in the night. Now we were certain that our abilities would be used to the fullest by our human counterparts.

Shee-na ended the Fifth World's prophecy with these words: "The Rainbow Warriors of the Fifth World of Illumination will be welcomed by the Stone Tribe as wholly healed Children of Earth. The Two-leggeds will no longer need to *Drop Their Robes,* or fear illness or death. They will be able to heal through their own thoughts, using the strengths of their own Medicine. Each Rainbow Tribe member will begin to use fully the best of his or her gifts without loss of memory. *The Remembering* will be a part of the World of Illumination and signals the beginning of all illusions being lifted."

The night had ended, and dawn brought the rays of Grandfather Sun's love, as if on cue. We reveled in the promised triumph of the Planetary Family and shared congratulations with the Little People and the Standing People who rimmed the edge of the cliffs. Even the Creepy-crawlers who had come from their nests to listen danced among the lights cast from the morning's dawning. The Creature-beings who had formed at the outer edge of our circle cried out in happy exclamations and then turned to share the news with others of their Clans down below Sacred Mountain, on the plains. There was a newfound joy and contentment present on Sacred Mountain, and it was good to be alive.

And so the prophecy of the Fifth World came into being.

CHAPTER EIGHT

THE SIXTH WORLD OF PROPHECY AND REVELATION

The morning had dawned, high on Sacred Mountain. Below in the valleys, plains, and as far west as the ocean shore the Creatures stirred, making ready for another day. The night had been long and filled with the revelation of the Fifth World Prophecy. Many of the Little People, having held vigil with our Council all through the night, were now sleeping.

I glanced around admiringly as I saw them nestled in groups, dozing amid the roots of Standing People, their faces peacefully content. The Wee-ones' determination to be of service to the Stone Clan had been proven time and again as they acted as emissaries or as our transportation during the progressive journey on Earth. They were also the Guardians of the Mineral Kingdom, making sure that our libraries of Earth Records were not unwittingly stolen by some human who sought material gain through selling a Gem Stone Person. I was grateful for the arms and legs of each tiny body that offered my Tribe a way to move, a way to be protected. Great Mystery had given the Rock Tribe a wonderful gift in these Brothers and Sisters. These Wee-ones were endowed with such an enormous sense of loyalty and a desire to be of service that their devotion never needed to be questioned.

This night would bring another Council Fire, when the prophecy of the Sixth World would be shared. I allowed my thoughts to turn to the coming evening and wondered how the story of the Sixth World would unfold. I had been a part of the group that carried the pieces of the Sixth World puzzle, and yet the import of the whole picture remained a mystery to me. I knew, as did all the others in my group, that the Sixth World was a world of prophecy and revelation. As I felt the significance of those two concepts, I began to form pictures in my mind. The prophecy that would come from the Sixth World was bound to be revealed to the Planetary Family as predictions of their own making. The concepts that were mentally created by All Our Relations, in that future time, would come as revelations of awareness. For millions of moons the Two-leggeds had called this idea *walking their talk,* and soon they would be *walking their thoughts.*

As this illuminating idea came to me, I was suddenly excited by potential possibilities. I began to understand how the Fifth World would prepare the Planetary Family for further growth. As all life-forms worked together as one during the Fifth World of Illumination, the stage would be set for All Our Relations to interact with one another, cocreating unified Councils and Tribes that would include all life-forms.

It was all I could do to imagine the Human Tribe understanding how it *felt* to be a Honey Bee or an Ocean Wave. Then I was suddenly hit with the idea that I would know how it felt to walk on legs like my Creature counterparts or how it would feel to have stems or limbs that would bud and blossom as a part of my body. My mind whirled in a million directions at once as I pondered the potentials that I could experience in the future.

The whirlwind of mental activity left me exhausted as I looked deeply into the pregnant Void of what would come. There, amid the chaos of my own expectation, stood many potentials that exploded into a wonder-filled set of new realities. Could the Planetary Family exceed even my limited understanding of the potentials Great Mystery had created before the beginning of time? I answered myself in the affirmative and grasped a stray thought that floated by. The planetary illumination of the Fifth World was going to remove all barriers to seeing and knowing all times, places, concepts, feelings, wisdom, and forms without separation. After that unification, every life-form would be able to learn and grow eternally without limitation. I reflected on this idea of limitless creation and was further astounded at the implications when I realized that all life-forms would truly know and feel everything that their counterparts felt and knew.

My thoughts crept back to the present state of being here in the Third World. I found that my earlier grief about the present state of affairs and the control of the Afraid of Dirts over the other races was greatly relieved. There was new hope and promise on the horizon, no matter how many fires humankind had to walk through in order to evolve spiritually.

I quieted myself and sent a thought of thanks to the Great Mystery for my enlightenment, and I watched as the shadows of early afternoon gently danced the dusk into being. I realized that life always held the quest for truth and the promise of realized visions, if we were only willing to feel it. Because of my morning revelation, I knew that one day all life-forms would be able to feel one another without separation. The joy of this promise sang in my heart and caused my spirit to soar.

That night we gathered together once more and watched the dancing flames of our Council Fire reach out into the starlit heavens as if to touch the heart of the Great Star Nation. Eebo the Elder, Chief of the Little People who had brought us safely to Sacred Mountain, spoke simply and eloquently as we opened our circle to hold Council. "Sisters and Brothers of the Stone Tribe, we of the Little People would like to assist you in bringing these messages of change to the Children of the Two-legged Tribe. We have held our own Council and have made decisions regarding our commitment to servicing unified growth on the Mother Planet.

"We are rarely ever seen by the Two-leggeds as they stray farther and farther from the natural world," Eebo continued. "Because of our invisibility, we would like to offer our services as a Give-away to the Earth Tribes. When any of you find a Two-legged who can be taught by a Stone Person, we will carry you to a place where you will be easily found by that human. In this way, we can be of further

service to both Clans. Those of the Human Clan will feel that they have found a Stone Teacher on their own, and you will be in the right place at the right time for them to find you."

A murmur of approval went around the circle as Eebo continued. "We Little People have found that sometimes we have to trick the Two-leggeds into thinking that an idea was all of their own making in order for their sense of self-importance to allow them to proceed," he chuckled. "If we are all going to work together to get those of the Human Tribe to see what is in front of their noses, we thought our assistance would be a welcome gift."

The exclamations of laughter and agreement filled the night as the Stone Tribe and the Little People decided to join forces. In convincing the Human Clan that they had created the wondrous future changes without any outside help, the Little People would count the ultimate coup. The Divine Trickster was certainly working some Coyote Medicine with Eebo and the others in coming up with that victorious idea. All present agreed that this plan would have a future punch line that would bring a lesson of humility through humor to the Two-leggeds when they could finally accept it. Eebo and the Little People were thanked, and then Shee-na, the Chief Stone Person, shared the tidings of the Sixth World.

"The Sixth World will be called the World of Prophecy and Revelation," Shee-na began. "This coming time will teach the Planetary Family how all vision is self-fulfilling prophecy. We currently understand that the combined thoughts of All Our Relations make up our common reality. In the Sixth World, however, we will discover that our personal revelations will create new prophecies, and those Dreamtime visions will immediately come into being without the influence of time.

"This is to say that life will develop through dream awareness." Shee-na smiled as she allowed this last concept to penetrate the thoughts of the various Clans who were listening. "Each life-form will contribute its dreaming to the unified dream-vision of the whole. The larger the dream, the faster the growth of the Planetary Family. Every part of Creation will dream the wisdom of every other part of Creation, allowing Knowing Systems to be birthed in the minds of All Relations.

"The Great Mystery will then be fully acknowledged as the Source, Creator of all realms of existence and the source of every physical manifestation." Shee-na paused for a moment and then explained, "An unbreakable connection to the Great Mystery will live in the hearts of us all.

"The secret of the Earth's magnetic pull will be revealed, as well as the effects that magnetism had on all of the preceding worlds. The function of other planets and stars will be revealed as our common understanding grows, allowing every other circle of life to contribute its wisdom and understanding to the whole." When Shee-na had completed the reading of the prophecy, she concluded with a final comment.

"My piece to the puzzle of the prophecy has to do with an overall truth, creating effects in all life-forms throughout all worlds. I want to share it with you now because it affects us in the Third World as much as it affects our future. In the First World, all that came into manifestation was a reflection of each individual reality or concept of *Self*. The First World reality was that all life-forms were made of love. For a time, each part of Creation lived through the idea of that unconditional love. Every life-form had a place of distinction that marked its unique talents and right to be. The creation of Brother Time gave all life-forms a constant rhythm through which all growth changes could pass. As we move through time, now in the Third World, we are given the opportunity to become the dream reflections of the whole. In our future, we will simply know and be." Shee-na's voice had taken on an almost shy quality as she ended, leading us in thoughts of praise and thanksgiving to the Original Source for the revelations that we had received.

The group broke into small bands as some sought the silence of the night. Others went to rest beneath the starry wonder of the Sky Nation. I sat amid my thoughts, marveling at the chain of events that had passed in front of the Council Fire this remarkable summer's eve. Tomorrow would bring the prophecy of the Seventh World. My thoughts wandered to the concept of time. I decided to dream through the remainder of the night, till dawn, offering to make Time my Ally because the hour had finally arrived for me to use my gift of wisdom in order to understand truth.

So it was that the future promises of the Sixth World awoke from the Earth Tribe's dream.

CHAPTER NINE

THE SEVENTH WORLD OF COMPLETION

Wisdom had visited me during the night as I traveled beyond the stars into realms I would have never before thought existed. In the dream, I met with Time and saw him standing with the Three Fates: Past, Present, and Future. None of these four identities had forms. I had expected them to appear as giant rock formations, which I later realized was a reflection of my own self-image. I had been fooled by my sense of self-importance into believing that spirits that I felt were more important than me must have forms like my own, but bigger.

Future was not the possessive type of being I had imagined. "Seven Talents," Future chuckled, "you are certainly learning the lessons of the Third World's control. We do not feel that anything *belongs* to any of the Three Fates. You will see when you arrive in the the Fifth World that every life-form will attain the ability to change its concept of Past, Present, and Future when linear time disappears."

"That's right," replied Time. "I don't even belong to myself. The linear part of me will vanish just like an old habit, and the unified whole of what I represent will appear in its place." I struggled to understand what Time had been trying to teach me, and then I asked, "Time, are you saying that just like the rest of Our Relations, you will become a part of the Uni-world?" "Of course," Time shouted, "You don't think that any part of Creation is ever wasted, do you?" I had never thought about it in that way, so I was shocked to see the holes that had just appeared in my rather faulty logic. Great Mystery certainly would not create anything that could be used only for a while and then must be discarded. It seemed natural that all Creation would just change form.

I was a bit timid after Time had shouted at me, but I continued. "Oh, Time, I came here to ask you to be my Ally. Now, I'm certain you can be a teacher for me but I'm not so sure you would like to be my Ally." Time began to snicker, then to laugh, then to guffaw, until the Three Fates took up a bow and shot energy arrows of hilarious laughter all over the Dreamtime. I started to get angry, but all four were laughing so hard, I began to laugh, too. I realized that they were laughing not at me but at the ridiculousness of the situation. I was so limited by my linear thoughts, from the expanded view they held, that my reasoning had tricked me into near idiocy.

Finally, Past got a hold on things while the other three continued to send waves of rocketing exuberance into deep space. "Now look here, Seven Talents," Past whistled, "Just where do you think all the old identities go when they are used up? Do you think they vanish forever with all of the energy it took every individual to create them?" Before Past allowed me to answer, Past's next statement was already whizzing through my mind. "Of course not! All of the old identities that any life-form ever used come to live with me, in the past. Just as soon as Time gives up the linear part of himself, the energy invested in all those old roles everyone acted out will be free for all of us to use, creating the Uni-world."

I felt as if I had been spanked for being naughty and for not thinking these concepts through on my own. I wasn't going to feel sorry for myself, so I made peace with the fact that I had decided to dream this crazy dream in order to make Time my Ally. I resigned myself to the Three Fates and chose my destiny with Time when I asked the fatal question, "Time, what do you need from me in order for you to become my Ally?"

The unbelievable laughter started again as Time urged the Three Fates to use a little more dignity and deportment. Time paused to give weight to his upcoming statement and then said, "Well, Seven Talents, you are the kind of Ally I would enjoy having, because you are a Record Keeper, and if I had use for anything, it would be for a historian of the events that passed through me. I guess that you would need to be meticulous about truth and you would have to be industrious, and of course, you would need to continue developing your seven greatest abilities." I was waiting for the other shoe to drop, but during the long pause I began to relax.

The Three Fates still giggled, while Time seemingly took a moment to allow me to make a decision. I nodded my agreement and Time continued. "There is another concept you might want to consider, Geeh Yuk. I am the Father of the Three Fates and they too will be dropping their old identities in the Fifth World to become one unified spirit along with me. Together we will become the *Ever-present Present*. In becoming my Ally, do you really think you can put up with this crazy trio of Fates? If not, we will have to part ways, because as each day passes, the four of us come closer to becoming one another. It's as if the original barriers that separated us were eroding into nothingness. Even the Two-leggeds can Stop the World when they Enter the Silence and see beyond the veils of forgetting."

I made my decision to become the Record Keeper and an Ally of Time. I promised to faithfully record all that happened on Earth and to put up with the silly siblings the Three Fates represented. Present chuckled and promptly proceeded to command my destiny for a moment by sneezing hard enough to blow me out of the Dreamtime and back into the *now* that existed on top of Sacred Mountain, Planet Earth. I was startled as I jolted back into my rock form. Night and day had passed and the Little People were adding wood to that night's Council Fire.

It was difficult to believe that I had been dreaming for that long. Perhaps if I had needed sleep, as did my Creature or Little People Relations, I would have been back sooner. It was then that I realized that there was no time in the Dreamtime. They had been laughing because I had carried my misconceptions through the crack in the universe to the place where all things were already unified and one. Time and the Three Fates had needed to manifest as their old identities in order to teach me about something that was in their past and yet still waiting to manifest in my future. I was shaken as I pieced together the journey that had brought me further understanding. I began to understand that Time was already my Ally and had gone to such trouble to help me learn. I roared with laughter as I saw the joke I had played on myself in order to make myself understand.

I reveled in the joy of my Dreamtime creation until it was time to approach the Council Fire, and then I took my place among the others. As I glanced around the circle, I realized that we created a giant Medicine Wheel with our bodies of stone. Outside our circle, the Little People were forming one of their own, and beyond that circle was the Council of Creature-beings. I longed in that moment to see the day when the Two-leggeds would be a part of our Council Fires, but my longing was not filled with grief as it would have been before. The prophecy of the worlds to come had given me a knowing that filled the emptiness of the unfulfilled dream of the future.

Shee-na once again took her place as our Chief and began the Council with words of thanksgiving for the guidance thus far provided. The hushed expectations of The Faithful permeated the air as the Wind Chief gently blew the scent of Evergreens into the circle surrounding our Council Fire. Shee-na began to speak the prophecy of the Seventh World of Completion, while All Relations thoughtfully listened to her words. "The Seventh World will allow all life to return to its beginning place. Each world passes through its phase of truth and lessons in order to evolve. Every Wheel of Life returns to its starting place and moves to the next spiral of the Sacred Dance of Life with greater wisdom and stronger Medicine. The lessons of the crooked trails, as well as the Pathways of Beauty along the way, bring truth and knowing. When everyone has stood on every spoke of the Great Medicine Wheel and understands the feelings of all other Relations, we will have reached completion. Then the Tribes of Earth will no longer need the lessons that cause separation, for life will be united as one.

"Since the Seventh World is the World of Completion, we will see the final refinements of all forms of communication. Every member of the Planetary Family will develop the Medicine of self-expression without hesitation or misunderstanding. This will mean that the languages of the Stones, Plants, Creatures, and Two-leggeds will be equally understood. The languages and feelings of the Little People, the Cloud People, the Winds of Change, and the Clan Chiefs of Air, Earth, Water, and Fire will be as easily accessible and understood as as one's personal thoughts. *Hail-lo-way-an*, the Language of Love, will return as the form of communication. This Love Language will be used as a means to understand the purest form of another's heart's desire. All individuals will be allowed to choose the method of self-expression they wish to use when communicating in *Hail-lo-way-an*. Some will choose dance, others art, others singing or feeling."

Shee-na paused and smiled as she spoke to the Council. "You see, all our languages will be fully understood by the Two-leggeds. I foresee great possibilities when we are able to exchange our ideas and truly serve one another. I know that the human children will be happy to hear the Creature Teachers and Little People again." Contented whispers of agreement were heard around the circle before she continued.

"All manifestations in the physical world will express the totality of their potentials, which will create constant communication between all parts of the whole. This in turn will cause limitless creation. Within the creative potential of the whole, we will see the unified Medicine of creativity, which will open new worlds of thought. The expansion of the Uni-world will constantly bring forth, complete and perfect, new parts to the whole. The spirit of this unified harmony will support the Planetary Family as one mind, one body, and one heart."

The silence was deafening as Shee-na continued. "The Whirling Rainbow of Peace will shower the Seventh World with Rainbow lights that will seal truth into the Vibral Cores of every life-form in the Uni-world. This essence of truth comes from Great Mystery and is manifested as the Eternal Flame. The Eternal Flame will remain intact throughout infinity. The Uni-world will become an identity of pure truth and love that will, as one unit, pulse the messages of wholeness and unified cocreation into Creation forever. Our Mother Planet will then take on a blanket of living light, creating a wall of fire around her body. Then the Earth Mother will become the second sun-star in this Wheel of Planets."

Whispers echoed through the crowd as we pictured the flaming light that would one day issue forth in spontaneous combustion from the darkness of the Void. Although it was improper to interrupt, I could not stop the question that sprang to my lips as I called out to Shee-na, "Will the Two-leggeds, Creature-beings, or Plant People perish from the flames or heat in the wall of fire?" Shee-na smiled as if she had anticipated a question of this type, and then she answered. "No, Seven Talents, they will already have bodies made of living light. When the wall of fire encircles the Earth Mother, every life-form will also become part of the Eternal Flame." A wave of relief passed through the encampment as we all felt our former concern and fear dissolve. Shee-na prepared to continue as the whispers subsided.

"The Seventh World of Completion will mark the return to our beginnings, as, in the First World, our sustenance came from the unconditional love of Grandfather Sun. When we complete the lessons of all prior worlds, we will not only learn how to return to that state of oneness but also become the living body of another sun."

When Shee-na had finished, we sat in awe trying to grasp the fullness of all that had passed in our three-day Council. The fourth day would be spent in celebration and thanksgiving. During our vigil we had discovered the potential for our Planetary Family, and our hearts had been lightened of the present burdens of the Third World. We knew that we would have to walk the paths of time until those prophecies were fulfilled. We decided to find joy in one thought. Along the way, we could share our good Medicine with others who might have lost hope as they encountered the crooked trails of their Earthwalks.

The Earth Mother had saved a final reward for her faithful Stone People, which came to us during our day of celebration. When we gave thanks for the blessings we had received as Record Keepers, we heard her voice on the Wind. "Children of my Stone Tribe, for your faithfulness, you will be permanently etched with the promise of the worlds to come. Your bodies will carry the symbols that will reveal the Pathways of Truth to those who seek your counsel. All beings who seek the Pathways of Truth may read the markings on the sides of your bodies, and they will be given the Medicine of wisdom they need to find their talents and to *Walk Their Talk*."

Cheers rang through the night as the Rock Tribe received the symbols of the Pathways of Truth. The markings were etched by the hands of the Lightning Beings and forged forever into our bodies. Every Stone Person carried the signs that would reflect each one's personal talents as well as the abilities of each to be of service to the Two-legged Tribe. The secrets of our Earth Records were set aglow deep within our hearts so that, through touch, the humans could grow with us in wisdom by feeling the prophecy held within.

Much time has passed since that fateful Council Fire in the Third World. The Children of Earth now stand at the dawning of the Fifth World of Illumination. The time is right for us to share the symbols of the Pathways of Truth, which can be found on the bodies of Stone People all over our planet.

These markings are gifts of prophecy, signposts of growth and the form of *Hail-lo-way-an*, the Language of Love, which is used by our Rock Tribe.

I, like you, am still refining my Seven Talents. The gift we share is growth. When we reach the Seventh World of Completion we will know the Unified World of Love that we once shared at the time of manifestation. There are many paths to be followed until then. Time is an Ally to us all. There will be plenty of time for each of us to complete our lessons of wholeness, and therefore, we need not have any fear. The Stone Tribe represents a reminder of the Earth Mother's love for all of her children. We offer our legends and our markings to those who seek *The Remembering* of truth as a shelter amidst the storm of lies and illusions. The Sacred Hoop remains unbroken because it is carried forever in the hearts of The Faithful.

Together, we are the promise of the worlds to come, and our mission is the same. We are here to develop our talents in order to complete our contribution of wholeness to life. Every spoke on the Medicine Wheel is sacred. The Pathways of Truth guide us to our common goal of completion as we each stand on every spoke of the Medicine Wheel of Life. The beauty of every lesson is invested with the Eternal Flame and its secret is understanding the Language of Love.

I have gone back to the Earth Mother to hold a new Council Fire like the one held during the Third World. This

**OTHER
COUNCIL
FIRES**

Council Fire will add to the former prophecies that mark the passage of the Planetary Family. As you walk on the surface of the Earth Mother, look into the faces of the Rock Tribe and seek the secrets of our new Council Fire. As the oldest of Earth Tribes, we are here to serve you, and our records are our gifts that will guide your paths. Know that we are eager to share those gifts . . . in the name of love.

LANGUAGE
OF THE STONES

BASIC SYMBOLS

Circles

The circle is the shape of harmony, representing perfection for Time Eternal. It is the symbol of the Creator, the Infinite Spirit, the Medicine Wheel, Sacred Space. In stone reading, the circle means a valuable lesson learned.

The circle is life, female, family; the extended family, sisters, brothers, aunts, uncles, and grandparents in all species; the identity of all creatures, plants, and elements; survival, or attunement to nature; the Universal World; oneness or wholeness; the idea of all for one, one for all.

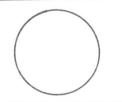 Large circle. The sun, the guardian of the day; the color yellow; the power of love; the male gender.

 Small circle. Changes and growth; the moon, the guardian of the night; the color white; the power of growth; lessons of change and growth; the power of perception.

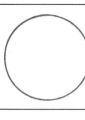

Perfect circle located at the center of a stone. The family, the female (family and female are synonymous in this instance); great powers of sensitivity in regard to changes.

Dots. Emphasize combinations; guidelines between connecting symbols. The single dot standing alone does not have a specific meaning.

Dot within a circle, at the center of the stone. A union, bond, or marriage; extended family; universal friendship. The dot within the circle is the seed, or the male. The circle is the female nourishing the seed.

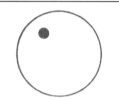

Dot within a circle, not in the center. Growth in that particular quadrant. (See final wheel illustration to discern the meanings of each direction or quadrant.)

The following symbols are in addition to the basic circles and give a more in-depth view of one's potential in relation to the family and the universe. These are powerful symbols. If they appear at the center of the stone, the individual who was attracted to this stone should carry it at all times. We call it a true environmental stone. These people are good listeners and have the ability to say the right thing at the right time, whenever the need arises. They are quiet, sincere, and dependable, and they sense where they can be helpful. Their intuitive sense of well-being makes it a joy to be in their company.

Circle within a circle. Relatives or relationships: the extended family, sisters, brothers, aunts (other mothers), uncles, grandparents; the power of insight.

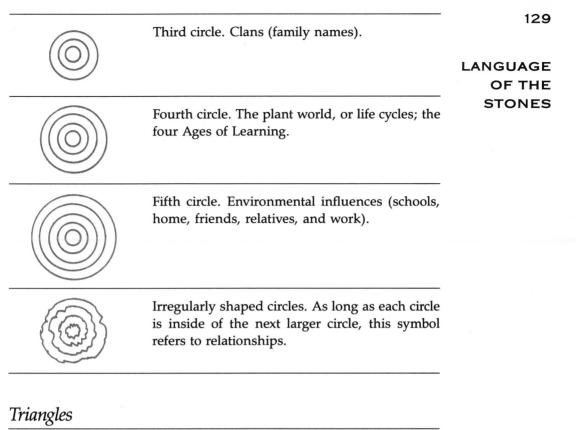

Third circle. Clans (family names).

Fourth circle. The plant world, or life cycles; the four Ages of Learning.

Fifth circle. Environmental influences (schools, home, friends, relatives, and work).

Irregularly shaped circles. As long as each circle is inside of the next larger circle, this symbol refers to relationships.

Triangles

Large equilateral triangle. Equality; wisdom; the gift of inner knowing; sensitivity.

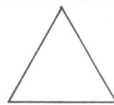

Small equilateral triangle. Gifts and talents; the power of creativity; latent talents to be developed.

Large inverted equilateral triangle. Confidence; the power of faith; recognizing a Sacred Point of View.

Small inverted equilateral triangle. Inspiration; the power of spirituality; a person who can inspire others.

Left closed right-angle triangle. Awareness or self-confidence; the power of awareness; accepting physical strength and gratitude.

Right closed right-angle triangle. Development of talents and gifts; the power of motivation.

Angles

Left right-angle. Learning; the power of honoring truth and wisdom; developing principles.

Right right-angle. Developing confidence in others; the power of work; being helpful to others. Working for the greatest good.

Left acute angle. The power of accepting responsibility; respect for someone or something. The ability to respond.

Right acute angle. The power of acceptance; accepting honor for achievements. Accepting the truth. Receiving graciously.

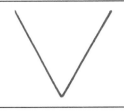

Upright V. A creative idea; the power of the mind; the Sacred Point of View. The ability to reason, think for oneself, and use the intellect positively.

Inverted V. Strength; the power of truth; a person who is not easily influenced by half-truths or glibness.

Squares

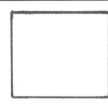

Square. Organization and protection; the power of stability; dependability. Foundations are firmly in place.

Diamonds

Diamond. Life, unity, equality for eternity; freedom from fear; the protective power of the winds; the Four Ages of Learning. (See final illustration.)

Rectangles

Vertical rectangle. Ability; the power of concentration, or centering in on a subject; self-sufficiency (sometimes, a loner or one who prefers to work alone). Focus and intent.

Horizontal rectangle. A variety of lessons in relation to one subject; the power of repetition; a person who needs to experience many lessons.

Parallelogram. Willingness to learn; the power of emergence; a person capable of accepting new ideas. If the ends lean left: achieving goals. If the ends lean right: becoming inspired.

Straight Lines

Vertical line. Energy; the power of spirit; a spiritual life path, moving upward; Earth Spirit.

Horizontal line. Fertility; the Earthwalk; the power of will; an earthy person.

Slanted lines. Directions. Leaning to the right: bringing learning gifts down to one's center. Leaning to the left: bringing magnetic powers down to one's center. Left and right slanted lines together: the power of leadership. Leading through example.

Curved Lines

Upturned curved line or U. Centers in on healing and gratitude in relation to spiritual growth. Growth achieved through healing old pain with thanksgiving.

Inverted U or dome. The power of influence; actions that result in feelings of trust and worthiness; lessons that build a foundation upon which to grow; a continuing process of growth.

U facing northwest. Points to the lessons that influence or provide a pattern for growth, for self as well as for others.

U facing northeast. Sets the principles that govern the pattern of growth. Learning about integrity and personal ethics through experience.

Dome facing southwest. Points to the emotions that serve the growth patterns of influence. Facing feelings in order to heal and grow.

Dome facing southeast. Testing influences and assessing them for personal growth. Adopting own truths by eliminating outside influences.

Wavy line. Energy; the use of energy on matter; physical activities affected by energy. Change created by emanations of energy.

COMBINATIONS OF BASIC SYMBOLS (MATERIAL)

Circle with straight lines. Sun with numerous rays; Medicine Wheel with twelve directions; reaching out for love and happiness.

Sun (convex—raised "bubble" on surface of stone) with seven rays. Talents and the ability to develop them; healing; good listener; peaceful person.

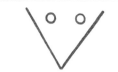

Dot-Sun (concave—indentation in stone) with four straight lines in the north, east, south, and west directions. Spiritual awareness; piety.

V under two small circles. Inquisitive person. Seeker of truth or the mysteries of life.

Inverted V over a circle. Talkative person. Ability to express ideas verbally.

Two small circles joined by a curved line on top. Hanging onto possessions; a collector of material objects. May need to learn how to Give Away.

Inverted V over a straight line. Closed mouth; quiet person; good listener.

X made up of two crossed lines or two angles meeting in the middle. Man; the male gender; the male power; masculinity; a workable idea.

X with a small circle between the upper arms. Woman; the female gender; the female power; nourishment. Compassion and nurturing.

Broken X. Ideas need refining. The Stone Person can teach you how to refine ideas and focus.

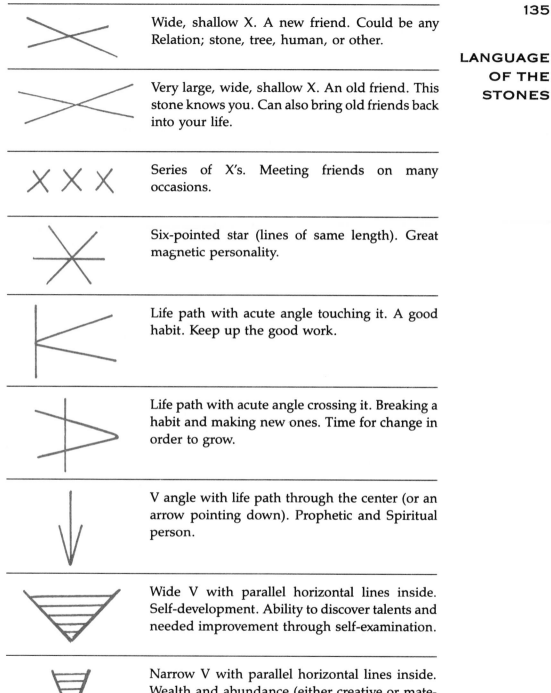

Wide, shallow X. A new friend. Could be any Relation; stone, tree, human, or other.

Very large, wide, shallow X. An old friend. This stone knows you. Can also bring old friends back into your life.

Series of X's. Meeting friends on many occasions.

Six-pointed star (lines of same length). Great magnetic personality.

Life path with acute angle touching it. A good habit. Keep up the good work.

Life path with acute angle crossing it. Breaking a habit and making new ones. Time for change in order to grow.

V angle with life path through the center (or an arrow pointing down). Prophetic and Spiritual person.

Wide V with parallel horizontal lines inside. Self-development. Ability to discover talents and needed improvement through self-examination.

Narrow V with parallel horizontal lines inside. Wealth and abundance (either creative or material). Ability to access this abundance.

V with curved ends. Person who can look at both sides of a question. A good or fair judge.

Vertical "arrow" with two lines on each side of the shaft, at the top. Many talents. Challenge may be to find and use them.

Vertical line with two lines on each side. Lasting friendship bond (marriage).

Vertical line with two lines coming at an angle out of the bottom. Spiritual strength (expressed in work—pointing to creativity).

Vertical line with two lines on each side, one set at the top and one in the middle. Creative ideas rising; linkage of ideas to life. In directions relates to the Medicine Wheel of Truth.

Check mark. Open to ideas of others (checking it out).

Two connected V's. Potential leader; one who sees and listens to others; the connection of the two points of view.

Two connected inverted V's. Strength in perception; believing in one's impressions. Trusting inner messages.

Wide U with an inverted V in center. Not easily influenced because of listening for positive expressions; one who doesn't gossip.

Wide U. Gift giving; a positive, faithful person who is devoted to discipline, speaking of the truth.

Wide U set on a horizontal line. A teacher of multiple talents, but not necessarily a master of them.

Wide U enclosed between two parallel, horizontal lines. One who is master of many talents; one who exhibits many talents but does not teach them.

Horizontal line at right angles to a vertical line and touching it. A decision made.

Horizontal line and vertical line at right angles, not touching. A decision not yet made; the ability to make good judgments.

Life path (vertical line) and earth path (horizontal line) at right angles, but with wide separation. Pending decisions requiring evaluations.

Life path and earth path meeting, with broken lines along the earth path. Facing many decisions.

"Cross," or horizontal and vertical lines crossing at right angles in the center. A stable, noncommittal person.

"Cross" with the vertical line not touching the horizontal line on either side. A person who prefers to be alone; a recluse.

Two parallel, horizontal lines crossed at right angles in the center by a vertical line. A very dependable person.

Two parallel, horizontal lines. A person faithful to any cause.

Wavering, horizontal line over and parallel to a solid line. A person who is easygoing one moment and indecisive the next.

Two parallel, wavering horizontal lines. A flexible person.

Straight, horizontal line with a curved line over and parallel to it and extending beyond the straight line at the right end. A person who needs encouragement from time to time.

Two parallel, horizontal lines with the right ends curved and crossing at the right. A person who relies upon others outside of the home. Ability to honor an extended family or Relations.

Two parallel, horizontal lines with the left ends curved and crossing at the left. A person who relies upon others within the home. Good family relationships.

Two parallel, vertical lines with the bottom ends curved and crossing at the bottom. A person who relies upon self. Trusting own abilities.

Two parallel, vertical lines with the top ends curved and crossing at the top. A person who relies upon spiritual concepts.

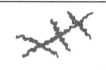 Broken path (two or more horizontal rows of short dashes.) Movement. Forward advancement.

Slanted, wavering line, moving up or down. Someone needs help.

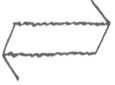 Slanted, wavering line with an angle of wavering lines touching it. You are helping someone in distress.

Several scattered, wavering lines crossed at an angle by another one. A time of self-evaluation.

 Two parallel, wavering, horizontal lines (not evenly one above the other, but positioned as in a parallelogram) with ends covered by angles. Sound, harmony, heard physically. Can change inner conflict to peace.

Two short arrows going in opposite directions, with shafts on same line. The power of direction in relation to desire. Can point out that you are working with or against your heart's desire.

Bird. Free spirit with high ideals.

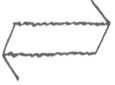

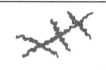

LANGUAGE OF THE STONES

These symbols represent the natural talents and abilities each Stone Person was given in the beginning, during Creation, and has learned throughout eons of time. When one or more of these symbols is found on your specific Stone Person, the lessons on how to master these talents is being offered to you.

Straight vertical line. The power of Spirit to conquer all challenges found in the material world.

Large X with two small, wide angles set in the east and west sides of the X, and two small arrows, pointing to the center, set in the north and south ends of the X. The power of eternity.

Connected, alternating light and dark circles and dashes. The influence of time. Making time, needing time, taking time, or altering time.

Many small vertical lines set at right angles on a straight, horizontal line. Life and life-force. Ability to use it well.

Medicine Wheel around the center of a cross, with its "arms" extended beyond the Medicine Wheel. The power of Four Ages: birth (growth), youth (adoption), adult (improvement), elders (wisdom). Cycles of growth and change.

Large circle (the sun symbol) with vertical diameter extending beyond the circle. The power of attraction. How to use it wisely.

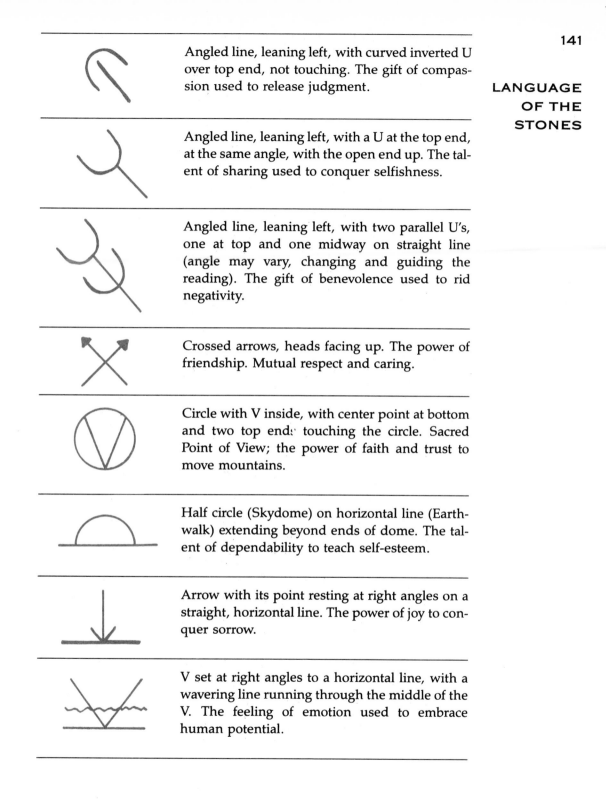

Angled line, leaning left, with curved inverted U over top end, not touching. The gift of compassion used to release judgment.

Angled line, leaning left, with a U at the top end, at the same angle, with the open end up. The talent of sharing used to conquer selfishness.

Angled line, leaning left, with two parallel U's, one at top and one midway on straight line (angle may vary, changing and guiding the reading). The gift of benevolence used to rid negativity.

Crossed arrows, heads facing up. The power of friendship. Mutual respect and caring.

Circle with V inside, with center point at bottom and two top ends touching the circle. Sacred Point of View; the power of faith and trust to move mountains.

Half circle (Skydome) on horizontal line (Earthwalk) extending beyond ends of dome. The talent of dependability to teach self-esteem.

Arrow with its point resting at right angles on a straight, horizontal line. The power of joy to conquer sorrow.

V set at right angles to a horizontal line, with a wavering line running through the middle of the V. The feeling of emotion used to embrace human potential.

Two V's, one above the other and parallel, with vertical line running through the center points. The power of illumination to teach spiritual clarity.

 Two parallel, vertical lines with two small circles on each at bottom and middle. The power of foresight used to strengthen and assure future success.

 Two angled, parallel lines, leaning left, with an angled "roof" at top end. The power of understanding used to promote awareness.

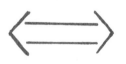 Two V's (W) on a horizontal line. The power of knowledge used to facilitate life.

 Snaking, wavering line, slanting, with arrow at bottom end. The power of wisdom. Facing left: self-understanding. Facing right: farsightedness.

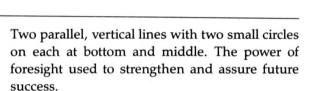 Two parallel, horizontal lines with angled "roofs" at each end, not touching the lines. The power of endurance used to meet challenges face to face.

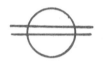 Circle with larger V standing up in it and extending above the top of the circle. The power of patience used to increase understanding.

Circle crossed by two extended, horizontal lines. The power of forgiveness used to heal the self and others.

V and half circle set at right angles to a horizontal line, with half circle bridging the point of the V. The power of healing in all its forms.

Vertical line with small circle touching at bottom. The power of mind used in a positive way to influence our world.

Circle with horizontal line through its center and extending beyond the sides of the circle. The power of growth and abundance used for the good of all.

Three or more connected V's on a horizontal line. The power of learning used to open new doors of opportunity.

Vertical line with a small V on either side of it, at bottom, set at right angles to a horizontal line. The gift of leadership. Ability to wisely use authority.

Horizontal line with short arrow, pointing up, above it but not touching it. The power of alertness. The ability to be present. Keen senses and talent of observation.

"Morning star": triangle with small, parallel marks on either side at the top, at right angles to the sides of the triangle. The power of guidance used to map the Path of Beauty.

Large X with small, horizontal line underneath in center. The power of influence used with wisdom, not for control.

	Two parallel V's one above the other. The talent of ambition used to become one's personal best.
	Horizontal arrow, pointing to the right. The power of inspiration used to enhance creativity.
	Horizontal arrow, pointing to the left. The power of direction used to hone desire into action.
	Short arrow, pointing down, set in the center of a wide, curved line, facing up. The power of discipline. Being a disciple of truth.
	Vertical line with a small circle at the bottom, resting at right angles on a horizontal line. The power of body; walking in harmony. Honoring the spirit's vehicle.
	Two parallel, horizontal lines, crossed at right angles by one vertical line. The power of strength to create physical and spiritual endurance or perserverance.
	Two parallel, vertical lines crossed by two parallel, horizontal lines, with the bottom horizontal line meeting the bottom ends of the vertical lines. The power of health. Honoring the body's needs.
	Small, parallel, vertical lines under a curved line. The power of bountiful food; rain.
	Small circle above a horizontal line. The power of sleep and rest to refresh and strengthen the body, mind, and spirit.

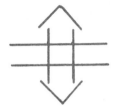

Horizontal and vertical lines in a tic-tac-toe pattern, with a V at the bottom, not touching the vertical lines, and an inverted V at the top, not touching them. The power of work to give purpose to life..

Vertical line (spirit) in center of a wide V (making an arrow), set inside a wide U but not touching it. The power of exercise to move energy and tone the body.

Half circle on a horizontal line, touching at each and (Skydome). Living: beginning at birth (Earthbirth), rising in spirit, and ending by giving over to death; the power of progress to bring wholeness and completion.

Arrow with wavering shaft, pointing to the right. The power of speed to access what is needed now.

Two parallel V's one above the other, with a narrow, vertical rectangle crossing their points. The power of fortitude to access valor, courage, and boldness.

Wide-angle zigzag, like a W, with its right "arm" on the bottom, pointing to the left. The power of location to bring a sense of well being and balance.

Two inverted, parallel V's, one above the other. The power of desire used to fire the creative abilities within.

Two linked U's overlapping. Sensitivity to opposite sex; the power of courtship used to grow closer.

Two small circles, side by side, with a short, vertical line touching the top of each circle. The power of sight used to develop the talent of observation.

Circle with a V on either side, near the top, with center points of the V touching the circle (like ears). The power of sound (inside psychic sounds) used to hear the messages of Spirit.

Two linked, overlapping U's above a horizontal line, but not touching it. The power of smell used to access awareness.

Inverted V over a small circle, both resting on a horizontal line. The power of taste used to taste life.

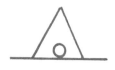

W set on a horizontal line with an inverted, curved line (C) above the inside center point of the W. The power of touch used to feel, sense, caress, and heal.

This symbol denotes using good habits for personal betterment. The power of good habits to maintain well being.

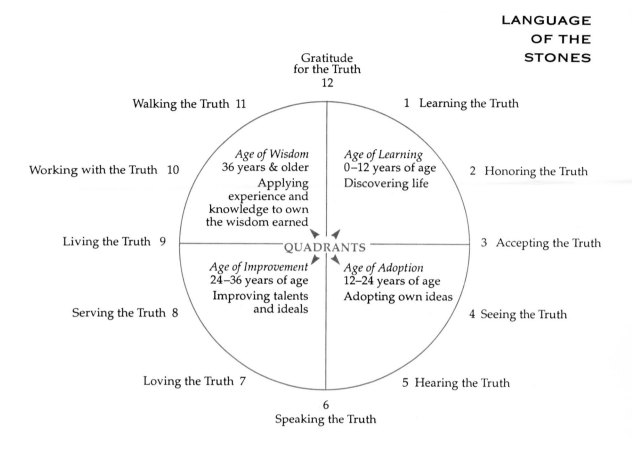

Gratitude
for the Truth
12

Walking the Truth 11

1 Learning the Truth

Age of Wisdom
36 years & older
Applying
experience and
knowledge to own
the wisdom earned

Age of Learning
0–12 years of age
Discovering life

Working with the Truth 10

2 Honoring the Truth

Living the Truth 9

QUADRANTS

3 Accepting the Truth

Age of Improvement
24–36 years of age
Improving talents
and ideals

Age of Adoption
12–24 years of age
Adopting own ideas

Serving the Truth 8

4 Seeing the Truth

Loving the Truth 7

5 Hearing the Truth

6
Speaking the Truth